Algorithms for
Discrete Fourier Transform
and Convolution

Richard Tolimieri   Myoung An   Chao Lu

Algorithms for Discrete Fourier Transform and Convolution

C.S. Burrus
Consulting Editor

Springer-Verlag
New York   Heidelberg   Berlin
London   Paris   Tokyo   Hong Kong

R. Tolimieri
Myoung An
Chao Lu
Center for Large Scale Computing
City University of New York
New York, NY 10036-8099, USA

Consulting Editor
Signal Processing and Digital Filtering

C.S. Burrus
Professor and Chairman
Department of Electrical and
    Computer Engineering
Rice University
Houston, TX 77251-1892
USA

Library of Congress Cataloging-in-Publication Data
Tolimieri, Richard, 1940–
    Algorithms for discrete fourier transform and
convolution.
    Includes bibliographical references.
    1. Fourier transformations—Data processing.
    2. Convolutions (Mathematics)—Data Processing.
    3. Digital filters (Mathematics) I. An, Myoung.
    II. Lu, Chao. III. Title.
QA403.5.T65 1989   515′.723   89–21692
ISBN 0-387-97118-1

Printed on acid-free paper

Camera-ready text prepared by the authors using T$_E$X.
Printed and bound by Braun-Brumfield, Inc., Ann Arbor, Michigan.
Printed in the United States of America.

9 8 7 6 5 4 3 2 1

ISBN 0-387-97118-1 Springer-Verlag New York Berlin Heidelberg
ISBN 3-540-97118-1 Springer-Verlag Berlin Heidelberg New York

PREFACE

This book is based on several courses taught during the last five years at the City College of the City University of New York and at Fudan University, Shanghai, China in the Summer, 1986. It was originally our intention to present to a mixed audience of electrical engineers, mathematicians and computer scientists at the graduate level, a collection of algorithms which would serve to represent the vast array of algorithms designed over the last twenty years for computing the finite Fourier transform (FFT) and finite convolution. However, it was soon apparent that the scope of the course had to be greatly expanded. For researchers interested in the design of new algorithms, a deeper understanding of the basic mathematical concepts underlying algorithm design was essential. At the same time, a large gap remained between the statement of an algorithm and the implementation of the algorithm. The main goal of this text is to describe tools which can serve both of these needs. In fact, it is our belief that certain mathematical ideas provide a natural language and culture for understanding, unifying and implementing a wide range of digital signal processing (DSP) algorithms. This belief is reenforced by the complex and time-consumming effort required to write code for recently available parallel and vector machines. A significant part of this text is devoted to establishing rules and precedures which reduce and at times automate this task.

In Chapter 1, a survey is given of basic algebra. It is assumed that much of this material is not new and in any case, the facts are easily described. The tensor product is introduced in chapter 2. The importance of the tensor product will be a reoccuring theme throughout this text. Tensor product factors have a direct interpretation as machine instructions on many vector and parallel computers. Tensor product identities provide linguistic rules for manipulating algorithms to match specific machine characteristics. Inherit in these rules are certain permutations, called stride permutations, which can be implemented on many machines by a single instruction. The tedious effort of readdressing, required in many DSP algorithms, is greatly reduced. Also, the data flow is highlighted which is especially important on super computers where the data flow is usually the major time-consumming part of the computation.

The design of fast DFT algorithms can be dated historically back to the times of Gauss (1805) [1]. The collected work of some unpublished manuscripts by Gauss contained the essentials of Cooley-Tukey FFT algorithm, but it did not attract much attention. Until 1965, when Cooley-Tukey published their paper [2], known as Fast Fourier Transform algorithm, the computing science started a revolutionary era. Since then, many variants of the Cooley-Tukey FFT algorithm have been developed. In chapter 3 and 4 of this book, the Cooley-Tukey FFT algorithm along with its many variants are unified under the banner of tensor product. From one point of view, these algorithms depend on mapping a one-dimensional array of data onto a multi-dimensional array of data (depending on the degree of compositeness of the transform size). Using tensor product we need derive only the simplist case of mapping into a 2-dimensional array. Tensor product identities can then be used to derive the general case. An explicit description of the data flow is automatically given along with rules for varying this data flow, if necessary. In chapter 5, the Good-Thomas PFA algorithm is reformulated by tensor product.

In chapter 6 and 7, various linear and cyclic convolution algorithms are described. The Chinese Remainder theorem (CRT) for polynomials is the major tool. Matrix and tensor product formulations are used wherever possible. Results of Cook-Toom and

Winograd are emphasized. The integer CRT is applied in chapter 7 to build large convolution algorithm from efficient small convolution algorithms (Agarwal-Cooley).

The scene changes in chapter 8. Various Multiplicative FFT algorithms (depending on the ring structure of the indexing set) are described. The prime size algorithms are due to Rader. Winograd generalized Rader's method to composite transform size. We emphasize a variant of the Rader-Winograd method. Tensor product language is used throughout and tensor product identities serve as powerful tools for obtaining several variants which offer arithmetic and data flow options.

In chapter 13, we consider the duality between periodic and decimated data established by the Fourier transform. This duality is applied in the computation of the Fourier transform of odd prime power, transform sizes, say p^k. The ring structure of the indexing set has an especially simple ideal structure (local ring). The main result decomposes the computation of the Fourier transform into two pieces. The first is Fourier transform of transform size p^{k-2}. The description of the second is the main objective of chapters 14 and 15, where we introduce the theory of multiplicative characters and derive formulas for computing the Fourier transform of multiplicative characters.

The author's are indebted to the paitience and knowledge of L. Auslander, J. Cooley and S. Winograd, who over the years at IBM, Yorktown Heights and at the City University of New York have taken time to explain their works and ideas. The authors wish to thank Professor C. S. Burrus, who read the manuscript of the book and suggested many improvements. A DARPA grant has given the authors the time to expand and test these ideas.

CONTENTS

Chapter 1. Introduction to Abstract Algebra

Chapter 2. Tensor Product and Stride Permutation

Chapter 3. Cooley-Tukey FFF Algorithms

Chapter 4. Variants of FFT Algorithms and Their Implementations

Chapter 5. Good-Thomas PFA

Chapter 6. Linear and Cyclic Convolution

Chapter 7. Agarwal-Cooley Convolution Algorithm

Chapter 11. <u>MFTA</u>: Transform Size $N = Mr$

M-Composite Integer and r-Prime

Chapter 12. <u>MFTA</u>: Transform Size $N = p^2$

Chapter 13. Periodization and Decimation

Chapter 14. Multiplicative Character and the FFT

Chapter 15. Rationality

Chapter 1

INTRODUCTION TO ABSTRACT ALGEBRA

1.1. Introduction

In this and the next chapters, we present several mathematical results needed to design the algorithms of the text. We assume that the reader has some knowledge of groups, rings and vector spaces but no extensive knowledge is required. Instead, we focus on those mathematical objects which will be used repeatedly in this text.

The ring of Integers Z

The ring Z/n of Integers $mod\, n$

Polynomial Rings $F[x]$

Quotient Polynomial Rings $F[x]/f(x)$

Tensor Products (chapter 2).

The *Chinese Remainder Theorem* for integers and polynomials will be discussed in detail as it plays a major role in many areas of algorithm design. The construction of the Chinese Remainder ring-isomorphism using *idempotents* has important implications to the rest of this work. This construction results in our algorithms having simple data flow.

The second tool which will play a major role is the tensor or Kronecker product of matrices and the commutivity theorem for tensor products. An algorithm will be described in many cases as a matrix factorization

$$X = X_1\, X_2\, \cdots\, X_r \qquad (1)$$

where the action of the matrix X is to be computed. This factorization decomposes the action of X into a sequence of 'elementary' operations

$$X_r,\, X_{r-1},\, \cdots,\, X_1. \qquad (2)$$

These elementary operations in many cases will be described using tensor products. The commutativity theorem gives a formal procedure for constructing variants of a particular factorization. These variants are significant since they can be matched to a given computer architecture. This is a theme which will recur throughout this work.

1.2. <u>The Ring of Integers</u>

The ring of integers Z satisfies the following important condition:

<u>Divisibility Condition</u> If a and b are integers with $b \neq 0$ then we can write

$$a \equiv bq + r, \qquad 0 \leq r < b, \tag{1}$$

where q and r are uniquely determined integers.

The integer q is called the *quotient* of the division of a by b, and it is the largest integer satisfying

$$bq \leq a. \tag{2}$$

The integer r is called the *remainder* of the division of a by b and is given by the formula

$$r = a - bq. \tag{3}$$

If $r = 0$ in (1), then

$$a = bq, \tag{4}$$

and we say that b *divides* a or that a is a *multiple* of b, and we write

$$b/a. \tag{5}$$

An integer $p > 1$ is called a *prime* if its only divisors are ± 1 and $\pm p$.

Consider integers a and b. An integer c is called a *common divisor* of a and b if c/a and c/b. The integers 1 and -1 are common divisors of any two integers a and b. If 1 and -1 are the only common divisors of a and b, we say that a and b are *relatively prime*. In any case, there are only a finite number of common divisors of two integers a and b as long as $a \neq 0$ or $b \neq 0$. Denote the largest common divisor of integers a and b by

$$(a, b). \qquad (6)$$

We call (a, b) the *greatest common divisor* of a and b. If $b = p$, a prime, then either $(a, p) = 1$ and a and p are relatively prime or $(a, p) = p$ and p/a.

Fix an integer n, and set

$$(n) = n\,Z = \{nk \mid k \in Z\}, \qquad (7)$$

the set of all multiples of n. The set $n\,Z$ is an *ideal* of the ring Z in the sense that it is closed under addition,

$$nk + nl = n(k + l), \qquad (8)$$

and closed under multiplication by Z,

$$m(nk) = (mn)k = n(mk), \qquad (9)$$

We will now prove a fundamental property of the ring Z.

Lemma 1. Every ideal of Z has the form $n\,Z$, for some integer $n \geq 0$.

Proof Suppose M is an ideal in Z. If $M = (0) = 0\,Z$, we are done. Otherwise, M contains positive integers and a smallest positive integer n. Take any $c \in M$ and write by (1),

$$c = nq + r, \qquad 0 \leq r < n.$$

Since c and n are in M,

$$r = c - nq$$

is also in M. However, $0 \le r < n$ which contradicts the definition of n unless $r = 0$. Thus $c = nq$ and $M = nZ$, proving the lemma.

We see from the proof of lemma 1 that any ideal $M \ne (0)$ in Z can be written as nZ where n is the smallest positive integer in M.

We will use lemma 1. to give an important description of the greatest common divisor.

<u>Lemma 2.</u> If a and b are nonzero integers, then

$$(a, b) = ax_0 + by_0, \tag{10}$$

for some integers x_0 and y_0.

Proof The set

$$M = \{ax + by \mid x, y \in Z\},$$

is an ideal of Z. By lemma 1,

$$M = dZ,$$

where d is the smallest positive integer in M. In particular, since a and b are in M, d is a common divisor of a and b. Now write

$$d = ax_0 + by_0,$$

and observe that any common divisor of a and b must divide d, proving the lemma.

From the proof of lemma 2, we have that every common divisor of a and b divides (a, b) which can also be characterized as the smallest positive integer in the set

$$M = \{ax + by \mid x, y \in Z\}. \tag{11}$$

As a consequence, $(a, b) = 1$ if and only if $ax_0 + by_0 = 1$, for some integers x_0 and y_0. We will use this to prove the following result.

Lemma 3. If a/bc and $(a, b) = 1$ then a/c. In particular, if p is a prime and p/bc then p/b or p/c.

Proof Since $(a, b) = 1$,

$$ax_0 + by_0 = 1,$$

for some integers x_0 and y_0. Then

$$c = cax_0 + cby_0.$$

Since a/cax_0, and by assumption a/bc, we have that a/c. To prove the second statement, we observe that if p does not divide b then $(p, b) = 1$. Applying the first part completes the proof of lemma 3.

We have all the ingredients needed to prove the fundamental *prime factorization* theorem.

Theorem 1. If $n > 1$ is an integer, then n can be written uniquely, up to ordering of the factors, as

$$n = p_1^{a_1} \cdots p_r^{a_r}, \tag{12}$$

where $p_1, \cdots, p_r$ are distinct primes and $a_1 > 0, \cdots, a_r > 0$ are integers.

Proof We first prove the existence of such a factorization. If n is prime, we are done. Otherwise write $n = n_1 n_2$ where $0 < n_1, n_2 < n$. By mathematical induction, assume n_1 and n_2 have factorizations of the form (12). Then their product $n = n_1 n_2$ can be written as a product of primes. Collecting the like primes, n has a factorization of the form (12). Suppose

$$n = q_1^{b_1} \cdots q_s^{b_s},$$

is a second factorization of the form (12). Then q_1/n. If $q_1 \neq p_j$, $1 \leq j \leq r$, then $(q_1, p_j) = 1$. Lemma 3 now implies q_1 does not divide

$p_j^{a_1}$. In this case, a second application of Lemma 3 implies q_1 does not divide n, a contradiction. It follows $q_1 = p_j$ for some $1 \leq j \leq r$. Continuing in this way, reordering the factors if necessary, we have $s \leq r$ and

$$q_k = p_k, \qquad 1 \leq k \leq s.$$

Reversing the roles of the prime factors, we have $s = r$ and

$$q_k = p_k, \qquad 1 \leq k \leq r.$$

Suppose $a_1 \leq b_1$. Applying the above argument to the integer

$$m = \frac{n}{p_1^{a_1}} = p_2^{a_2} \cdots p_r^{a_r} = p_1^{b_1 - a_1} p_2^{b_2} \cdots p_r^{b_r}$$

we have $b_1 = a_1$. Continuing in this way, uniqueness follows, completing the proof of the theorem.

Take an integer $n > 1$ and a prime p dividing n. Suppose a is the largest positive integer satisfying

$$p^a / n. \tag{13}$$

By the proof of theorem 1, p^a appears in the prime factorization of n. If now q is another prime divisor of n, and b is the largest positive integer satisfying

$$q^b / n, \tag{14}$$

then $p^a q^b$ appears in the prime factorization of n. This discussion leads to the following corollary of theorem 1.

Corollary If a/c and b/c and $(a, b) = 1$ then

$$ab/c. \tag{15}$$

Proof Since $(a, b) = 1$, the prime factors of a and b are distinct. Repeated application of the above discussion proves the corollary.

1.3. The Ring Z/n

Fix an integer $n > 1$. For any integer a, set

$$a \bmod n, \tag{1}$$

equal to the remainder of the division of n into a. In particular,

$$0 \leq (a \bmod n) < n. \tag{2}$$

Set

$$Z/n = \{0, 1, 2, \cdots, n-1\}. \tag{3}$$

Define addition in Z/n by

$$(a + b) \bmod n, \quad a, b \in Z/n, \tag{4}$$

and multiplication in Z/n by

$$(a \cdot b) \bmod n, \quad a, b \in Z/n. \tag{5}$$

Straightforward computation shows that Z/n becomes a commutative ring with identity 1 under these operations.

Consider the mapping

$$\eta : Z \longrightarrow Z/n, \tag{6}$$

defined by

$$\eta(a) = a \bmod n. \tag{7}$$

The mapping η is a *ring-homomorphism* in the sense that

$$\eta(a + b) = (\eta(a) + \eta(b)) \bmod n, \tag{8}$$

$$\eta(a \cdot b) = (\eta(a) \cdot \eta(b)) \bmod n. \tag{9}$$

Two integers a and b are said to be *congruent mod n* if $\eta(a) = \eta(b)$, or equivalently

$$n/(a - b). \tag{10}$$

In this case we write

$$a \equiv b \; mod \; n, \tag{10}$$

The *unit group* of Z/n, denoted by $U(n)$, consists of all elements $a \in Z/n$ which have multiplicative inverses $b \in Z/n$:

$$1 = (ab) \; mod \; n. \tag{12}$$

To show that $a \in U(n)$, it suffices to find an element $b \in Z$ such that

$$1 \equiv ab \; mod \; n, \tag{13}$$

since it then follows that

$$1 = (a(b \; mod \; n)) \; mod \; n. \tag{14}$$

Straightforward verification shows that $U(n)$ is a group under the ring-multiplication in Z/n.

Example 1. Take $n = 9$. Then

$$U(9) = \{1, 2, 4, 5, 7, 8\}.$$

The group table of $U(9)$ under multiplication *mod* 9 is as follows:

	1 2 4 5 7 8
1	1 2 4 5 7 8
2	2 4 8 1 5 7
4	4 8 7 2 1 5
5	5 1 2 7 8 4
7	7 5 1 8 4 2
8	8 7 5 4 2 1

Lemma 1. $U(n) = \{a \in Z/n \mid (a, n) = 1\}$.

Proof By remarks following lemma 2.2, $(a, n) = 1$ if and only if

$$a x_0 + n y_0 = 1,$$

for some integers x_0 and y_0. Equivalently $(a, n) = 1$ if and only if

$$ax_0 \equiv 1 \ mod \ n,$$

which by (13) implies $x_0 \ mod \ n$ is the multiplicative inverse of a in Z/n, proving the lemma.

Example 2. Take $n = 15$. Then

$$U(15) = \{1, \ 2, \ 4, \ 7, \ 8, \ 11, \ 13, \ 14\}.$$

As a special case of lemma 1, if p is a prime, then

$$U(p) = \{1, \ 2, \ \cdots \ , \ p - 1\},$$

and every nonzero element in Z/p has a multiplicative inverse. Since Z/p is a commutative ring with identity, it follows that Z/p is a finite field.

Lemma 2. Z/p is a finite field if and only if p is a prime.

Proof We have shown that if p is a prime then Z/p is a field. Suppose n is not a prime, and write $n = n_1 n_2$ where

$$1 < n_1, \ n_2 < n.$$

By lemma 1, since $(n, n_1) = n_1 \neq 1$, n_1 does not have a multiplicative inverse in Z/n and Z/n is not a field, completing the proof of lemma 2.

1.4. Chinese Remainder Theorem (CRT) I.

Suppose $n = n_1 n_2$ where $(n_1, n_2) = 1$. Form the *ring direct product*

$$Z/n_1 \ \times \ Z/n_2. \tag{1}$$

A typical element in (1) is an ordered pair

$$(a_1, a_2), \quad a_1 \in Z/n_1, \ a_2 \in Z/n_2. \tag{2}$$

Addition and multiplication in (1) are taken as componentwise addition

$$(a_1, a_2) + (b_1, b_2) = ((a_1 + b_1) \bmod n_1, \ (a_2 + b_2) \bmod n_2), \tag{3}$$

and componentwise multiplication

$$(a_1, a_2)(b_1, b_2) = ((a_1 b_1) \bmod n_1, \ (a_2 b_2) \bmod n_2). \tag{4}$$

The (CRT) constructs a ring-isomorphism

$$Z/n \cong Z/n_1 \times Z/n_2. \tag{5}$$

We will construct the ring-isomorphism using *idempotents*. Since $(n_1, n_2) = 1$,

$$n_1 f_1 + n_2 f_2 = 1, \quad f_1, \ f_2 \in Z. \tag{6}$$

Set

$$e_1 = (n_2 f_2) \bmod n, \tag{7}$$

$$e_2 = (n_1 f_1) \bmod n. \tag{8}$$

Rewrite (7) as

$$e_1 = n_2 f_2 + nq, \quad q \in Z. \tag{9}$$

We see from (6) that

$$e_1 \equiv 1 \bmod n_1, \tag{10}$$

and from (9) that

$$e_1 \equiv 0 \bmod n_2. \tag{11}$$

In the same way

$$e_2 \equiv 0 \bmod n_1, \tag{12}$$

$$e_2 \equiv 1 \bmod n_2. \tag{13}$$

The element $e_1 \in Z/n$ is uniquely determined by conditions (10) and (11). Suppose a second element $f_1 \in Z/n$ can be found satisfying

$$f_1 \equiv 1 \bmod n_1, \quad f_1 \equiv 0 \bmod n_2. \tag{14}$$

Then $f_1 \equiv e_1 \bmod n_1$ and $f_1 \equiv e_1 \bmod n_2$, implying

$$n_1/f_1 - e_1 \quad n_2/f_1 - e_1, \tag{15}$$

Since $(n_1, n_2) = 1$, the corollary to theorem 3.1 implies

$$n = n_1 n_2/f_1 - e_1. \tag{16}$$

Without loss of generality, we assume $f_1 - e_1 \geq 0$. We have

$$0 \leq f_1 - e_1 < n \tag{17}$$

which in light of (16) implies $f_1 = e_1$. This same argument shows that if a and b are elements in Z/n satisfying

$$a \equiv \bmod n_1, \quad a \equiv b \bmod n_2, \tag{18}$$

where $n = n_1 n_2$ and $(n_1, n_2) = 1$. This implies

$$a = b. \tag{19}$$

Conditions (10)-(13) uniquely determine the set

$$\{e_1, e_2\}, \tag{20}$$

which we call the *system of idempotents* for the ring Z/n corresponding to the factorization $n = n_1 n_2$, $(n_1, n_2) = 1$.

Table 1.   Examples of Idempotents.

n	n_1	n_2	e_1	e_2
6	2	3	3	4
10	2	5	5	6
12	3	4	4	9
15	3	5	10	6
21	3	7	7	15
28	4	7	21	8

The system of idempotents given in (20) will be used to define a ring-isomorphism (5). First we need the following result.

Lemma 1. If $\{e_1, e_2\}$ is the system of idempotents for Z/n corresponding to the factorization $n = n_1 n_2$, $(n_1, n_2) = 1$, then

$$e_1^2 \equiv e_1 \bmod n, \quad e_2^2 \equiv e_2 \bmod n, \tag{21}$$

$$e_1 e_2 \equiv 0 \bmod n, \tag{22}$$

$$e_1 + e_2 \equiv 1 \bmod n. \tag{23}$$

Proof By (10) and (11), $n_1/(e_1 - 1)$ and n_2/e_1. Thus

$$n_1/e_1(e_1 - 1), \quad n_2/e_1(e_1 - 1).$$

Since $(n_1, n_2) = 1$, and $e_1^2 - e_1 = e_1(e_1 - 1)$,

$$n = n_1 n_2 / e_1^2 - e_1,$$

proving
$$e_1^2 = e_1 \bmod n.$$

In the same way, $e_2^2 \equiv e_2 \bmod n$, proving (21). Since n_1/e_1 and n_2/e_2, $(n_1, n_2) = 1$ implies $n = n_1 n_2 / e_1 e_2$ and

$$e_1 e_2 \equiv 0 \bmod n.$$

Finally, $n_1/e_1 - 1$ and n_1/e_2, implying $n_1/e_1 + e_2 - 1$. In the same way $n_2/e_1 + e_2 - 1$. Again, $(n_1, n_2) = 1$ implies $n/e_1 + e_2 - 1$ and

$$e_1 + e_2 \equiv 1 \bmod n,$$

completing the proof of lemma 1.

Define the mapping

$$\phi : Z/n_1 \times Z/n_2 \longrightarrow Z/n, \tag{24}$$

by the formula

$$\phi(a_1, a_2) = (a_1 e_1 + a_2 e_2) \bmod n, \quad a_1 \in Z/n_1, \quad a_2 \in Z/n_2. \quad (25)$$

Theorem 1. ϕ is a ring-isomorphism of $Z/n_1 \times Z/n_2$ onto Z/n.

Proof Take

$$a = (a_1, a_2), \ b = (b_1, b_2) \in Z/n_1 \times Z/n_2.$$

We will write addition and multiplication in $Z/n_1 \times Z/n_2$ by $a + b$ and $a \cdot b$. Straightforward verification shows that

$$\phi(a + b) = (\phi(a) + \phi(b)) \bmod n.$$

Lemma 1 implies

$$\phi(a \cdot b) = (\phi(a)\phi(b)) \bmod n.$$

By (21) and (22)

$$\phi(a)\phi(b) \equiv a_1 b_1 e_1^2 + (a_1 b_2 + a_2 b_1) e_1 e_2 + a_2 b_2 e_2^2$$

$$\equiv a_1 b_1 e_1 + a_2 b_2 e_2 \bmod n.$$

Formula (23) implies

$$\phi(1, 1) \equiv 1 \bmod n,$$

proving that ϕ is a ring-homomorphism.

To prove that ϕ is onto, take any $k \in Z/n$ and observe that

$$k \equiv (k \bmod n_1) e_1 + (k \bmod n_2) e_2 \bmod n$$

Since $Z/n_1 \times Z/n_2$ and Z/n have the same number of elements, this proves that ϕ is onto, completing the proof of theorem 1.

From the above proof, we see that the inverse ϕ^{-1} of ϕ is given by

$$\phi^{-1}(k) = (k \bmod n_1, k \bmod n_2), \quad k \in Z/n. \tag{26}$$

This implies that every $k \in Z/n$ can be written uniquely as

$$k \equiv k_1 e_1 + k_2 e_2 \bmod n, \tag{27}$$

where $k_1 \in Z/n_1$ and $k_2 \in Z/n_2$. This fact will be used later.

Example 4.   Take $n = 15$ with $n_1 = 3$ and $n_2 = 5$. Then $e_1 = 10$ and $e_2 = 6$ and ϕ is given by the table below.

Table 2.

$Z/3 \times Z/5$	$Z/15$
(0,0)	0
(0,1)	6
(0,2)	12
(0,3)	3
(0,4)	9
(1,0)	10
(1,1)	1
(1,2)	7
(1,3)	13
(1,4)	4
(2,0)	5
(2,1)	11
(2,2)	2
(2,3)	8
(2,4)	14

Consider the *direct product of unit groups*

$$U(n_1) \times U(n_2), \tag{28}$$

with componentwise multiplication. $U(n_1) \times U(n_2)$ is the unit group of the ring $Z/n_1 \times Z/n_2$. In general, any ring-isomorphism maps the unit group isomorphically onto the unit group. Thus, we have the following theorem.

Theorem 2.   The ring-isomorphism ϕ restricts to a group-isomorphism of $U(n_1) \times U(n_2)$ onto $U(n)$.

The extension of theorem 1 to factorization,

$$n = n_1 n_2 \cdots n_r, \tag{29}$$

where the factors n_1, n_2, $\cdots$, n_r are pairwise relatively prime will be given in problems 8 to 12.

1.5. Unit Groups

Properties of unit groups play a major role in algorithm design. In this section, we will state, at times without proof, several important results which will be used repeatedly throughout the text.

Denote the number of elements in a set S by $o(S)$ and call $o(S)$ the order of S. In section 3, we proved

$$o(U(p)) = p - 1, \qquad \text{for a prime } p. \tag{1}$$

The same argument, using Lemma 3.1, proves for a prime p,

$$o(U(p^a)) = p^a - p^{a-1} = p^{a-1}(p - 1), \quad a \geq 1. \tag{2}$$

CRT, especially theorem 2., can be used to extend these results to the general case. Suppose

$$n = p_1^{a_1} \cdots p_r^{a_r}, \tag{3}$$

is the prime factorization n. Then

$$U(n) \cong U(p_1^{a_1}) \times U(p_2^{a_2}) \times \cdots \times U(p_r^{a_r}). \tag{4}$$

It follows that

$$o(U(n)) = p_1^{a_1-1} \cdots p_r^{a_r-1} (p_1 - 1) \cdots (p_r - 1). \qquad (5)$$

The function $o(U(n))$, $n \geq 1$, is called the *Euler quotient function*.

Table 1. Values of the Euler quotient function

n	$o(U(n))$
5	4
5^2	20
5^3	100
7	6
7^2	42
7^3	294
35	24

We require the following results from general group theory.

Theorem 1. If G is a finite group of order m with composition law written multiplicatively, then for all $x \in G$.

$$x^m = 1, \qquad (6)$$

Applying this result to the unit group of the finite field Z/p, we have from (1)

$$x^{p-1} \equiv 1 \bmod p, \qquad x \in U(p). \qquad (7)$$

Equivalently

$$x^p \equiv x \bmod p, \qquad x \in Z/p. \qquad (8)$$

Similar results hold in the unit groups $U(p^a)$ and $U(n)$.

The next two results are deeper and will be presented without proof.

Theorem 2. For an odd prime p, and integer $a \geq 1$, the unit group

$$U(p^a) \tag{9}$$

is a cyclic group.

This important result is proved in many number theory books, for instance [1]. As a consequence of theorem 1, an element $z \in U(p^a)$, called a *generator*, can be found such that

$$U(p^a) = \{z^k \mid 0 \leq k < o(U(p^a))\}. \tag{10}$$

The corresponding result for $p = 2$ is slightly more complicated. The unit group

$$U(2^2) = \{1, 3\} \tag{11}$$

is cyclic, but $U(2^a)$, $a > 2$, is never cyclic. The exact result follows.

Theorem 3. The group

$$U(2^a), \qquad a \geq 3 \tag{12}$$

is the direct product of two cyclic groups, one of order 2, the other of order 2^{a-2}. In fact, if

$$G_1 = \{1, -1\}, \tag{13}$$

$$G_2 = \{5^k \mid 0 \leq k < 2^{a-2}\}, \tag{14}$$

then

$$U(2^a) = G_1 \times G_2. \tag{15}$$

Example 5. For $p = 3$ and $a = 2$,

$$U(3^2) = \{2^k \mid 0 \leq k < 6\}.$$

Example 6. Take $p = 2$ and $a = 3$. Then

$$U(2^3) = \{1, 3, 5, 7\} = \{1, 7\} \times \{1, 5\}.$$

Take $a = 4$. Then

$$U(2^4) = \{1, 15\} \times \{1, 5, 9, 13\}.$$

1.6. <u>Polynomial Rings</u>

Consider a field F and denote by $F[x]$ the ring of polynomials in the indeterminate x, having coefficients in F. A typical element in $F[x]$ is a mathematical expression

$$f(x) = \sum_{k=0}^{r} f_k x^k, \quad f_k \in F. \tag{1}$$

If $f_r \neq 0$ in (1), we say that the *degree* of $f(x)$ is r and write

$$deg\ f(x) = r. \tag{2}$$

The elements of F can be viewed as polynomials over F. The non-zero elements in F can be identified with the polynomials over F of degree 0. The zero polynomial, denoted by 0, has by convention degree $-\infty$. Then, we have the important result

$$deg\ (f(x)\ g(x)) = deg\ f(x) + deg\ g(x). \tag{3}$$

The integer ring Z and polynomial rings over fields have many properties in common. The reason for this is that the following divisibility condition holds in $F[x]$.

<u>Divisibility condition</u> If $f(x)$ and $g(x) \neq 0$ are polynomials in $F[x]$, then there is a unique pair of polynomials $q(x)$ and $r(x)$ in $F[x]$ satisfying

$$f(x) = q(x)\ g(x) + r(x) \tag{4}$$

and

$$deg\ r(x) < deg\ g(x). \tag{5}$$

The polynomial $q(x)$ is called the *quotient* of the division of $g(x)$ into $f(x)$. In practice, we compute $q(x)$ by long division of polynomials. The polynomial $r(x)$ is called the *remainder* of the division of $f(x)$ by $g(x)$.

If $r(x) = 0$ in (4), we have

$$f(x) = q(x)g(x), \qquad q(x) \in F[x], \tag{6}$$

and we say that $g(x)$ *divides* $f(x)$ or $f(x)$ is a *multiple* of $g(x)$ over F, and we write

$$g(x)/f(x). \tag{7}$$

The elements of F viewed as polynomials are called *constant polynomials*. Nonzero constant polynomials divide any polynomial. A nonconstant polynomial $p(x)$ in $F[x]$ is said to be *irreducible* or *prime* over F if the only divisors of $p(x)$ in $F[x]$ are constants or a constant times $p(x)$. Constant polynomials play the same role in $F[x]$ that the integers 1 and -1 play in Z. To force uniqueness in statements below, we require the notion of a *monic polynomial* which is defined by the property that the largest degree nonzero coefficient is equal to 1.

The division relation satisfies the properties:

$$\text{If } h(x)/g(x) \text{ and } g(x)/f(x), \text{ then } h(x)/f(x). \tag{8}$$

$$\text{If } h(x)/f(x) \text{ and } h(x)/g(x), \text{ then } h(x)/(a(x)f(x) + b(x)g(x)), \tag{9}$$

for all polynomial $a(x)$ and $b(x)$.

$$\text{If } f(x)/g(x) \text{ and } g(x)/f(x), \text{ then } f(x) = a\, g(x), \ a \in F. \tag{10}$$

Consider polynomials $f(x)$ and $g(x)$ over F. A polynomial $h(x)$ over F is called a *common divisor* of $f(x)$ and $g(x)$ over F if

$$h(x)/f(x) \text{ and } h(x)/g(x). \tag{11}$$

We say that $f(x)$ and $g(x)$ are *relatively prime* over F if the only divisors of both $f(x)$ and $g(x)$ over F are the constant polynomials.

A subset J of $F[x]$ is called an ideal if J satisfies the following two properties:

$$\text{If } f(x), g(x) \in J, \text{ then } f(x) + g(x) \in J, \tag{12}$$

$$\text{If } f(x) \in J \text{ and } a(x) \in F[x], \text{ then } a(x)f(x) \in J. \tag{13}$$

Equivalently, J is an ideal if for any two polynomials $f(x)$ and $g(x)$ in J,

$$a(x)f(x) \; + \; b(x)g(x) \in J, \tag{14}$$

for all polynomials $a(x)$ and $b(x)$ in $F[x]$. The set

$$(f(x)) \; = \; \{a(x)f(x) \mid a(x) \in F[x]\} \tag{15}$$

is an ideal of $F[x]$. The divisibility condition will be used to show that all ideals J in $F[x]$ are of this form. The proof is the same as that in section 2., where we now use the divisibility condition for polynomials. First note that if an ideal J contains nonzero constants, then

$$J \; = \; (1) \; = \; F[x], \tag{16}$$

since by (13), if $a \neq 0$ is in J, then

$$f(x) \; = \; (a^{-1}f(x))a \tag{17}$$

is in J for arbitrary $f(x)$ in $F[x]$.

Lemma 1. If J is an ideal, other than (0) or $F[x]$, in $F[x]$, then

$$J \; = \; (d(x)), \tag{18}$$

where $d(x)$ is uniquely determined as the monic polynomial of lowest positive degree in J.

Proof By (13), J contains a monic polynomial of lowest <u>positive</u> degree, say $d(x)$. Take any $f(x)$ in J and write

$$f(x) = q(x)\, d(x) + r(x),$$

where $deg\ r(x) < deg\ d(x)$. By (14),

$$r(x) = f(x) - q(x)d(x),$$

is in J. Since $deg\ r(x) < deg\ d(x)$, we must have

$$deg\ r(x) = -\infty \text{ or } 0.$$

But J contains no nonzero constants, implying $r(x) = 0$. Since $f(x)$ is arbitrary,

$$J = (d(x)),$$

and all polynomials in J are multiples of $d(x)$. By (10), $d(x)$ is uniquely determined as the lowest positive degree monic polynomial in J, proving the Lemma.

Take any two polynomials $f(x)$ and $g(x)$ over F. The set

$$J = \{a(x)f(x) + b(x)g(x) \mid a(x),\ b(x) \in F[x]\}, \qquad (19)$$

is an ideal in $F[x]$. By Lemma 1,

$$J = (d(x)), \qquad (20)$$

where $d(x)$ is the monic polynomial of lowest degree in J. In particular, $d(x)$ is a common divisor of $f(x)$ and $g(x)$.

Write

$$d(x) = a_0(x)f(x) + b_0(x)g(x), \quad a_0(x),\ b_0(x) \in F[x]. \qquad (21)$$

By (9), every common divisor of $f(x)$ and $g(x)$ divides $d(x)$. We have proved the following result.

Lemma 2. If $f(x)$ and $g(x)$ are polynomials over F, then there exists a unique monic polynomial $d(x)$ over F satisfying

I. $d(x)$ is a common divisor of $f(x)$ and $g(x)$.

II. Every divisor of $f(x)$ and $g(x)$ in $F[x]$ divides $d(x)$.   (22)

Equivalently, $d(x)$ is the unique monic polynomial over F which is a common divisor of $f(x)$ and $g(x)$ of maximal degree. We call $d(x)$ the *greatest common divisor* of $f(x)$ and $g(x)$ over F and write

$$d(x) = (f(x), g(x)).   (23)$$

By the divisibility condition above

$$(f(x), g(x)) = a(x)f(x) + b(x)g(x),$$

where $a(x)$ and $b(x)$ are polynomials over F. In particular, if $f(x)$ and $g(x)$ are relatively prime over F, then

$$1 = a_0(x)f(x) + b_0(x)g(x),   (24)$$

for some polynomials $a_0(x)$ and $b_0(x)$ over F.

Arguing as in section 2, we have the following corresponding results.

Lemma 3. If $f(x)/g(x)h(x)$, $(f(x), g(x)) = 1$, then $f(x)/h(x)$.

Theorem 1. (Unique Factorization) If $f(x)$ is a polynomial over F then $f(x)$ can be written uniquely, up to an ordering of factors, as

$$f(x) = ap_1^{a_1}(x) \cdots p_r^{a_r}(x),   (25)$$

where $a \in F$, $p_1(x), \cdots, p_r(x)$ are irreducible polynomials over F and $a_1 > 0, \cdots, a_r > 0$ are integers.

Corollary   For polynomials over F, if $f(x)/g(x)$, $h(x)/g(x)$ and $(f(x), h(x)) = 1$, then

$$f(x)h(x)/g(x).$$

1.7. Field Extension

Suppose K is a field and F is a *subfield* of K in the sense that F is a subset of K containing 1, and it is closed under the addition and multiplication in K. For example, the rational field $\mathbf{Q}$ is a subfield of the real field $\mathbf{R}$ which is a subfield of the complex field $\mathbf{C}$ We also say K is an extension of F. Observe

$$F[x] \subset K[x]. \tag{1}$$

A polynomial $p(x)$ in $F[x]$ can be irreducible over F without being irreducible as a polynomial in $K[x]$. For example, the polynomial

$$x^2 + 1 \tag{2}$$

is irreducible over the real field, but over the complex field;

$$x^2 + 1 = (x + i)(x - i). \tag{3}$$

Thus, reference to the field of definition is necessary when dealing with irreduciblity.

Consider now the greatest common divisor $d(x)$ over F of two polynomials $f(x)$ and $g(x)$ over F. View $f(x)$ and $g(x)$ as polynomials over K and denote the greatest common divisor of $f(x)$ and $g(x)$ over K by $e(x)$. By Lemma 6.2,

$$d(x)/e(x), \tag{4}$$

meaning

$$e(x) = q(x)d(x), \qquad q(x) \equiv K[x]. \tag{5}$$

Write

$$d(x) = a(x)f(x) + b(x)g(x), \qquad a(x),\ b(x) \in F[x]. \tag{6}$$

Since $e(x)$ is a common divisor of $f(x)$ and $g(x)$, by (6.9),

$$e(x)/d(x). \tag{7}$$

Applying (6.10),

$$e(x) = d(x). \tag{8}$$

Thus greatest common divisor $d(x)$ of two polynomials $f(x)$ and $g(x)$ over F does not change when we go to an extension K of F. In particular, $f(x)$ and $g(x)$ are relatively prime over F if and only if they are relatively prime over any extension K of F.

Consider a polynomial $f(x)$ over F and suppose K is an extension of F. The polynomial $f(x)$ can be evaluated at any element a of the field K by replacing the indeterminate x by a. The result

$$f(a) = \sum_{k=0}^{r} f_k a^k \tag{9}$$

is an element in K. We say that a is a root or zero of $f(x)$ if $f(a) = 0$. The main reasons to consider extensions K of a field F is to find roots of polynomials $f(x)$. For instance, $x^2 + 1$ has no roots in the real field but in the complex field i and $-i$ are roots.

Lemma 1. A nonzero polynomial $f(x)$ over F has a root a in an extension field K if and only if

$$f(x) = (x - a)q(x), \tag{10}$$

for some polynomial $q(x)$ over K. In any extension field K of F, a polynomial $f(x)$ over F has at most n roots, $n = deg\ f(x)$.

Proof Applying the divisibility condition in $K[x]$

$$f(x) = (x - a)\ q(x) + r(x),$$

where $q(x), r(x) \in K[x]$ and $deg\ r(x) < 1$. Since $f(a) = 0$, we have $r(a) = 0$ implying $r(x) = 0$.

Suppose $a_1, \cdots, a_m$ are distinct roots of $f(x)$ in K. Then $(x - a_j)/f(x)$. The polynomials

$$x - a_1, \cdots, x - a_m,$$

are relatively prime. By the extension of the corollary of section 6. to several factors, we have

$$(x - a_1) \cdots (x - a_m)/f(x).$$

Thus

$$m \leq deg\ f(x),$$

completing the proof.

1.8. The Ring $F[x]/f(x)$

Fix a polynomial $f(x)$ over $F[x]$ of degree n. Set

$$F[x]/f(x), \tag{1}$$

equal to the set of all polynomials $g(x)$ over F satisfying

$$deg\ g(x) < n. \tag{2}$$

Every polynomial $g(x)$ in $F[x]/f(x)$ can be written as

$$g(x) = \sum_{k=0}^{n-1} g_k x^k, \qquad g_k \in F, \tag{3}$$

and we can regard $F[x]/f(x)$ as an n-dimensional vector space over F having basis

$$1,\ x,\ \cdots\ x^{n-1}. \tag{4}$$

We place a *ring multiplication* on $F[x]/f(x)$ as follows. For any $g(x) \in F[x]$, denote by

$$g(x)\ mod\ f(x), \tag{5}$$

the remainder of the division of $g(x)$ by $f(x)$. Then

$$g(x)\ mod\ f(x) \in F[x]/f(x). \tag{6}$$

Define multiplication in $F[x]/f(x)$ by

$$(g(x)\, h(x))\; mod\; f(x), \qquad g(x),\, h(x) \in F[x]/f(x). \tag{7}$$

Direct computation shows that the vector space $F[x]/f(x)$ becomes an *algebra* over F with the multiplication (7).

Two polynomials $g(x)$ and $h(x)$ over F are said to be congruent *mod* $f(x)$, and we write

$$g(x) \equiv h(x)\; mod\; f(x), \tag{8}$$

if $g(x)\; mod\; f(x) = h(x)\; mod\; f(x)$. Equivalently, (8) holds if

$$f(x)/(g(x) - h(x)). \tag{9}$$

Define the mapping

$$\eta \quad : \; F[x] \longrightarrow F[x]/f(x), \tag{10}$$

by the formula

$$\eta(g(x)) = g(x)\; mod\; f(x). \tag{11}$$

Straightforward computation shows that η is a ring-homomorphism of $F[x]$ onto $F[x]/f(x)$ whose kernel

$$\{g(x) \in F[x] \mid \eta(g(x)) = 0\} \tag{12}$$

is the ideal $(f(x))$.

In Lemma 6.2, we gave a method of constructing finite field of order p, for a prime p.

$$Z/p \text{ is a field if and only if } p \text{ is a prime.} \tag{13}$$

We will now construct fields using the rings $F[x]/f(x)$.

Lemma 1. The ring $F[x]/f(x)$ is a field if and only if $f(x)$ is irreducible over F.

Proof Suppose $f(x)$ is irreducible. Take any nonzero polynomial, $g(x)$ in $F[x]/f(x)$. By (6.24),

$$1 = a_0(x)g(x) + b_0(x)f(x),$$

where $a_0(x)$ and $b_0(x)$ are polynomials over F. Then

$$1 \equiv a_0(x)g(x) \bmod f(x),$$

so $a_0(x) \bmod f(x)$ is the multiplicative inverse of $g(x)$ in $F[x]/f(x)$. Since $g(x)$ is an arbitrary nonzero polynomial in $F[x]/f(x)$, the commutative ring $F[x]/f(x)$ is a field.

Conversely, suppose $f(x)$ is not irreducible. Then

$$f(x) = f_1(x)f_2(x),$$

where

$$0 < deg\ f_k(x) < deg\ f(x), \quad k = 1, 2.$$

Then, $f_1(x)$ and $f_2(x)$ are in $F[x]/f(x)$ and

$$0 = (f_1(x)f_2(x)) \bmod f(x).$$

If $f_1(x)$ has a multiplicative inverse, then

$$0 \equiv f_2(x) \bmod f(x),$$

a contradiction, completing the proof of the converse of the lemma.

More generally, we have the next result which we give without proof.

Lemma 2. The unit group U of $F[x]/f(x)$, consisting of all polynomials $g(x)$ in $F[x]/f(x)$ having multiplicative inverse is

$$U = \{h(x) \in F[x]/f(x) \mid (h(x), f(x)) = 1\}. \qquad (14)$$

Identifying F with the constant polynomials in $F[x]/f(x)$, we have

$$F \subset F[x]/f(x). \tag{15}$$

If $p(x)$ is an irreducible polynomial of degree n, then

$$K = F[x]/p(x). \tag{16}$$

is a field extension of F which can also be viewed as a vector space of dimension n over F. Suppose

$$F = Z/p, \tag{17}$$

then K is a finite field of order p^n. We state without proof the next result.

Lemma 3. If K is a finite field then K has order p^n for some prime p and integer $n \geq 1$. Two finite fields of the same order are isomorphic.

In addition, every finite field K can be constructed as in (16).

1.9. CRT for Polynomial Rings

Consider a polynomial $f(x)$ over F, and suppose

$$f(x) = f_1(x)f_2(x), \qquad (f_1(x), f_2(x)) = 1. \tag{1}$$

We will define a ring-isomorphism

$$F[x]/f_1(x) \times F[x]/f_2(x) \cong F[x]/f(x), \tag{2}$$

where the *ring-direct product* in (2) is taken with respect to componentwise addition and multiplication. First, following section 4, we define the *idempotents*. Since $f_1(x)$ and $f_2(x)$ are relatively prime, we can write

$$1 = a_1(x)f_1(x) + a_2(x)f_2(x), \tag{3}$$

with $a_1(x)$ and $a_2(x)$ polynomials over F. Set

$$e_1(x) = (a_2(x) \, f_2(x)) \, mod \, f(x), \tag{4}$$

$$e_2(x) = (a_1(x) \, f_1(x)) \, mod \, f(x). \tag{5}$$

Arguing as in section 4.,

$$e_1(x) \equiv 1 \, mod \, f_1(x), \quad e_1(x) \equiv 0 \, mod \, f_2(x), \tag{7}$$

$$e_2(x) \equiv 0 \, mod \, f_1(x), \quad e_2(x) \equiv 1 \, mod \, f_2(x). \tag{8}$$

Conditions (7) and (8) uniquely determined the polynomials $e_1(x)$ and $e_2(x)$ in $F[x]/f(x)$. The set

$$\{e_1(x), \, e_2(x)\}, \tag{9}$$

is called the *system of idempotents* corresponding to factorization (1).

Arguing as in Lemma 4.1, we have the next result.

Lemma 1. The system of idempotents (9) satisfies

$$e_k^2(x) \equiv e_k(x) \, mod \, f(x), \qquad k = 1, 2, \tag{10}$$

$$e_1(x)e_2(x) \equiv 0 \, mod \, f(x), \tag{11}$$

$$e_1(x) + e_2(x) \equiv 1 \, mod \, f(x). \tag{12}$$

Define

$$\phi(g_1(x), \, g_2(x)) = (\, g_1(x) \, e_1(x) + g_2(x) \, e_2(x)) \, mod \, f(x), \tag{13}$$

where $g_k(x) \in F[x]/f_k(x)$, $k = 1, 2$. As in theorem 4.1, the next result follows.

Theorem 1. ϕ is a ring-isomorphism of the ring-direct product $F[x]/f_1(x) \times F[x]/f_2(x)$ onto $F[x]/f(x)$ having inverse ϕ^{-1} given by the formula

$$\phi^{-1}(g(x)) = (g(x) \, mod \, f_1(x), \, g(x) \, mod \, f_2(x)), \tag{14}$$

for $g(x) \in F[x]/f(x)$.

In particular, every $g(x)$ in $F[x]/f(x)$ can be uniquely written as

$$g(x) \equiv g_1(x)\, e_1(x) + g_2(x)\, e_2(x) \; mod \; f(x), \tag{15}$$

where $g_k(x) \in F[x]/f_k(x)$, $k = 1, 2$.

The extension of these results to factorization of the form

$$f(x) = f_1(x)\, f_2(x) \, \cdots \, f_r(x), \tag{16}$$

where the factors $f_k(x)$, $1 \leq k \leq r$, are pairwise relatively prime, is straightforward. To construct the system of idempotents,

$$\{e_k(x) \mid 1 \leq k \leq r\}, \tag{17}$$

corresponding to the factorization (16), we reason as follows. First

$$(f_1(x), \; f_2(x) \, \cdots \, f_r(x)\,) = 1, \tag{18}$$

and we can apply the above discussion to find a unique polynomial $e_1(x)$ in $F[x]/f(x)$ satisfying

$$e_1(x) \equiv 1 \; mod \; f_1(x), \tag{19}$$

$$e_1(x) \equiv 0 \; mod \; f_2(x) \cdots f_r(x). \tag{20}$$

Condition (20) implies

$$e_1(x) \equiv 0 \; mod \; f_k(x), \qquad 1 \leq k \leq r, \; k \neq 1. \tag{21}$$

Continuing in this way, we find polynomials $e_k(x)$ in $F[x]/f(x)$, $1 \leq k \leq r$, satisfying

$$e_k(x) \equiv 1 \; mod \; f_k(x), \qquad 1 \leq k \leq r, \tag{22}$$

$$e_k(x) \equiv 0 \; mod \; f_k(x), \qquad 1 \leq k, l \leq r, \quad k \neq l. \tag{23}$$

These conditions uniquely determine the set (17). As before, we have the next result.

Lemma 2. The system of idempotents (17) satisfies the properties

$$e_k^2(x) \equiv e_k(x) \bmod f(x), \qquad 1 \le k \le r, \tag{24}$$

$$e_k(x)\, e_k(x) \equiv 0 \bmod f(x), \qquad 1 \le k, l \le r, \quad k \ne l, \tag{25}$$

$$\sum_{k=1}^{r} e_k(x) \equiv 1 \bmod f(x). \tag{26}$$

From Lemma 2, we have the ring-isomorphism ϕ of the direct product

$$\Pi_{k=1}^{r} F[x]/f_k(x) \tag{27}$$

onto $F[x]/f(x)$ given by the formula

$$\phi(g_1(x), \cdots, g_r(x)) = \left(\sum_{k=1}^{r} g_k(x)\, e_k(x) \right) \bmod f(x). \tag{28}$$

The inverse ϕ^{-1} of ϕ is given by the formula

$$\phi^{-1}(g(x)) = (g(x) \bmod f_1(x), \cdots, g(x) \bmod f_r(x)), \tag{29}$$

and every $g(x)$ in $F[x]/f(x)$ can be written uniquely as

$$g(x) \equiv \sum_{k=1}^{r} g_k(x)\, e_k(x) \bmod f(x), \tag{30}$$

where $g_k(x)$ is in $F[x]/f_k(x)$, $1 \le k \le r$.

[References]

[1] Ireland and Rosen *A Classical Introduction to Modern Number Theory*, Springer-Verlag 1980.

[2] Halmos, P. R. *Finite-Dimensional Vector Spaces*, Springer-Verlag 1974.

[3] Herstein, I. N. *Topics in Algebra*, XEROX College Publishing, 1964.

[References of Preface]

[1] Heideman, M. T., Johnson, D. H. and Burrus, C. S. "Gauss and the History of the Fast Fourier Transform", *IEEE ASSP Magazine*, October 1984.

[2] Cooley, J. W. and Tukey, J. W. "An Algorithm for the Machine Calculation of Complex Fourier Series", *Math. Comp.*, vol. 19, No. 2.

Problems

1. Show

$$(a + (b + c) \bmod n) \bmod n = ((a + b) \bmod n + c) \bmod n.$$

This is the associative law for addition *mod n*.

2. Show

$$(a \cdot ((b + c) \bmod n)) \bmod n$$

$$= ((a \cdot b) \bmod n + (a \cdot c) \bmod n) \bmod n.$$

This is the distributative law in the ring Z/n.

3. Describe explicitly the unit group $U(n)$ of Z/n for $n = 12$, $n = 21$, $n = 44$ and $n = 105$.

4. Give the table for addition and multiplication in the field $Z/11$.

5. Give the table for addition and multiplication in the ring $Z/21$.

6. Find the system of idempotents corresponding to the factorizations $n = 3 \times 7$, $n = 4 \times 5$, $n = 2 \cdot 7$ and $n = 7 \times 11$.

7. Give the table for the ring-isomorphism ϕ of the CRT corresponding to factorizations $n = 4 \times 5$ and $n = 2 \times 7$.

8. Suppose $n = n_1 n_2 \cdots n_r$ where the factors $n_1 n_2, \cdots, n_r$ are relatively prime in pairs. Show that

$$(n_k, n/n_k) = 1, \qquad 1 \leq k \leq r.$$

9. Continuing the notation and using the result of problem 8, define integers $e_1, e_2, \cdots, e_r$ satisfying

$$0 \leq e_k < n, \quad 1 \leq k \leq r,$$

$$e_k \equiv 1 \bmod n_k, \qquad 1 \leq k \leq r,$$

$$e_k \equiv 0 \bmod n_l, \qquad 1 \leq k, l < r, \ k \neq l.$$

These integers are uniquely determined by these conditions and form the system of idempotents corresponding to the factorization given in Problem 8.

10. Continuing the notation of problems 8 and 9, prove the analog of Lemma 1.

$$e_k^2 \equiv e_k \bmod n, \qquad 1 \leq k \leq r,$$

$$e_k e_l \equiv 0 \bmod n, \qquad 1 \leq k, l \leq r, \ k \neq l,$$

$$\sum_{k=1}^{r} e_k \equiv 1 \bmod n.$$

11. Define the Chinese Remainder ring-isomorphism ϕ of the direct product $Z/n_1 \times Z/n_2 \times \cdots \times Z/n_r$ onto Z/n and describe its inverse ϕ^{-1}.

12. Extend Theorem 2. to the case of several factors given in the above problems.

13. Find a generator of the unit group $U(n)$ of Z/n where $n = 5$, $n = 25$, $n = 125$.

14. Show that $U(21)$ is not a cyclic group. Use Theorem 2. to find generators of $U(21)$.

15. If $n = p_1 p_2 \cdots p_r$ where the factors $p_1, p_2, \cdots, p_r$ are distinct primes, show that the unit group $U(n)$ is
  group-isomorphic to the direct sum

$$Z/p_1 - 1 \oplus Z/p_2 - 1 \oplus \cdots \oplus Z/p_r - 1.$$

16. Prove the formula (6.3).

$$deg\ (f(x)\,g(x)) = deg\ f(x) + deg\ g(x).$$

17. Write out the divisibility condition for the polynomials

$$g(x) = x^{10} + 4x^8 + 2x^2 + 3,$$

$$f(x) = 4x^{20} + 2x^{10} + 1.$$

18. For any two polynomials $f(x)$ and $g(x)$ over $Q[x]$, show that the following set is an ideal,

$$J = \{a(x)f(x) + b(x)g(x) \mid a(x), b(x) \in Q[x]\}.$$

19. Let F be a finite field and form the set

$$L = \{1,\ 1+1,\ \cdots,\ 1+1+\cdots+1,\ \cdots\}$$

Show that L has order p for some prime p and that L is a subfield of F isomorphic to the field Z/p. The prime p is called the *characteristic* of the finite field F.

20. Show that every finite field K has order p^n for some prime p and integer $n \geq 1$.

21. If $f(x)$ is the polynomial over $\mathbf{Q}$

$$f(x) = (x-1)(x+1),$$

find the idempotents corresponding to this factorization and describe the table giving the CRT ring-isomorphism.

22. Find the idempotents corresponding to the factorization

$$f(x) = (x - a_1)(x - a_2) \cdots (x - a_r),$$

where $a_1, \cdots, a_r$ are elements in some field F. Describe the corresponding CRT ring-isomorphism ϕ and its inverse ϕ^{-1}.

Chapter 2

TENSOR PRODUCT AND STRIDE PERMUTATION

2.1. Introduction

Tensor product offers a natural language for expressing digital signal processing(DSP) algorithms. In this chapter, we define the tensor product and derive several important tensor product identities.

Closely associated with tensor products are a class of permutations, the stride permutations. These permutations govern the addressing between the stages of the tensor product decompositions of DSP algorithms. As we will see in the following chapters, these permutations distinguish the variants of the Cooley-Tukey FFT algorithms and other DSP algorithms.

Tensor product formulation of DSP algorithms also offers the convenience of modifying the algorithms to adapt to specific computer architectures. Tensor product identities can be used in the process of automating the implementation of the algorithms on these architectures. The formalism of tensor product notation can be used to keep track of the complicated index calculation needed in implementing Fourier transform (FT) algorithms. In [1], the implementation of tensor product actions on the CRAY XMP was carried out in detail.

2.2. Tensor Product

In this section, we present some of the basic properties of tensor products which are encountered in the algorithms that we will describe in future sections of this work. Tensor product algebra is an important tool for presenting mathematical formulations of DSP algorithms so that these algorithms may be studied and analyzed in a unified format. We first define the tensor product of vectors and present some of its properties. We then define the tensor product of matrices and describe additional properties. These properties will be very useful in manipulating the factorizations of discrete FT matrices.

Consider vectors $\underline{a}$ and $\underline{b}$ of sizes M and N, respectively. We write $\underline{a}$ and $\underline{b}$ as column vectors,

$$\underline{a} = \begin{bmatrix} a_0 \\ a_1 \\ \cdot \\ \cdot \\ \cdot \\ a_{M-1} \end{bmatrix}, \quad \underline{b} = \begin{bmatrix} b_0 \\ b_1 \\ \cdot \\ \cdot \\ \cdot \\ b_{N-1} \end{bmatrix}. \tag{1}$$

The *tensor product* $\underline{a} \otimes \underline{b}$ is the vector of size MN defined by

$$\begin{bmatrix} a_0 \underline{b} \\ a_1 \underline{b} \\ \cdot \\ \cdot \\ \cdot \\ a_{M-1}\underline{b} \end{bmatrix}. \tag{2}$$

For example,

$$\begin{bmatrix} a_0 \\ a_1 \\ a_2 \end{bmatrix} \otimes \begin{bmatrix} b_0 \\ b_1 \end{bmatrix} = \begin{bmatrix} a_0 b_0 \\ a_0 b_1 \\ a_1 b_0 \\ a_1 b_1 \\ a_2 b_0 \\ a_2 b_1 \end{bmatrix}. \tag{3}$$

The tensor product is *bilinear* in the following sense.

$$(\underline{a} + \underline{b}) \otimes \underline{c} = \underline{a} \otimes \underline{c} + \underline{b} \otimes \underline{c}, \tag{4}$$

$$\underline{a} \otimes (\underline{b} + \underline{c}) = \underline{a} \otimes \underline{b} + \underline{a} \otimes \underline{c}, \tag{5}$$

but it is not commutative. In general, $\underline{a} \otimes \underline{b} \neq \underline{b} \otimes \underline{a}$.

We can view tensor product formulation as relating a linear array to multidimensional arrays. An $M \times N$ array

$$X = [x_{m,n}]_{0 \leq m < M, 0 \leq n < N},   \tag{6}$$

will be identified with the vector $\underline{x}$ of size MN formed by running down the columns of X in consecutive order. Conversely, given a vector $\underline{x}$ of size MN we can form on $M \times N$ array by first *segmenting* $\underline{x}$ into N vectors of size M,

$$X_0 = \begin{bmatrix} x_0 \\ x_1 \\ \cdot \\ \cdot \\ \cdot \\ x_{M-1} \end{bmatrix}, \; X_1 = \begin{bmatrix} x_M \\ x_{M+1} \\ \cdot \\ \cdot \\ \cdot \\ x_{2M-1} \end{bmatrix}, \; \cdots,$$

$$X_{N-1} = \begin{bmatrix} x_{(N-1)M} \\ \cdot \\ \cdot \\ \cdot \\ x_{MN-1} \end{bmatrix}$$

and then placing these N vectors in N consecutive columns,

$$X = [X_0 \quad X_1 \quad \cdots \quad X_{N-1}].   \tag{7}$$

We may represent this procedure in a diagram as follows.

$$\underline{x} \xrightarrow{\text{segmentation}} \begin{bmatrix} X_0 \\ X_1 \\ \cdot \\ \cdot \\ \cdot \\ X_{M-1} \end{bmatrix} \xrightarrow{\text{array}} [X_0 \quad X_1 \quad \cdots \quad X_{N-1}].$$

Consider the tensor products $\underline{a} \otimes \underline{b}$ and $\underline{b} \otimes \underline{a}$. Identify $\underline{a} \otimes \underline{b}$ with the $N \times M$ array

$$X = [a_0 \underline{b} \quad \cdots \quad a_{M-1}\underline{b}],   \tag{8}$$

and $\underline{b} \otimes \underline{a}$ with the $M \times N$ array

$$Y = [b_0\underline{a} \quad \cdots \quad b_{N-1}\underline{a}]. \tag{9}$$

We see that $Y = X^t$. Thus interchanging order in the tensor product corresponds to matrix transpose. For example, the vector in (3) corresponds to the 2x3 array

$$\begin{bmatrix} a_0 b_0 & a_1 b_0 & a_2 b_0 \\ a_0 b_1 & a_1 b_1 & a_2 b_1 \end{bmatrix}, \tag{10}$$

while the vector $\underline{b} \otimes \underline{a}$ corresponds to the 3x2 array

$$\begin{bmatrix} b_0 a_0 & b_1 a_0 \\ b_0 a_1 & b_1 a_1 \\ b_0 a_2 & b_1 a_2 \end{bmatrix}. \tag{11}$$

In general, we can describe matrix transposition in terms of a permutation of an indexing set. Consider first the $N \times M$ array

$$Y = [y_{nm}]_{0 \le n < N, 0 \le m < M}. \tag{12}$$

Each r, $0 \le r < MN$, can be written uniquely as $r = n + mN$, with $0 \le n < N$, $0 \le m < M$. Also, each s, $0 \le s < MN$ can be written uniquely as $s = m + nM$, with $0 \le m < M, 0 \le n < N$. The vector $\underline{y}$ formed from the array Y has components given by

$$y_r = y_{nm}, \quad r = n + mN, \quad 0 \le m < M, \quad 0 \le n < N, \tag{13}$$

while the vector $\underline{z}$ formed from Y^t has components given by

$$z_s = y_{nm}, \quad s = m + nM, \quad 0 \le m < M, \quad 0 \le n < N. \tag{14}$$

This corresponds to the permutation of the indexing set,

$$\pi(n + mN) = m + nM, \quad 0 \le m < M, \quad 0 \le n < N, \tag{15}$$

and we have

$$y_r = z_{\pi(r)}, \quad 0 \le r < MN. \tag{16}$$

Example 1.   Taking $M = 2$ and $N = 3$, we have

$$\pi = (\, 0\ 2\ 4\ 1\ 3\ 5\,),$$

and

$$\underline{y} = \begin{bmatrix} z_0 \\ z_2 \\ z_4 \\ z_1 \\ z_3 \\ z_5 \end{bmatrix}.$$

To form $\underline{y}$ from $\underline{z}$ we 'stride' through $\underline{z}$ with length two.

In general, to form $\underline{a} \otimes \underline{b}$ from $\underline{b} \otimes \underline{a}$ we first initialize at $b_0 a_0$, the 0-th component of $\underline{b} \otimes \underline{a}$, then stride through $\underline{b} \otimes \underline{a}$ with length the size of $\underline{a}$. After a pass through $\underline{b} \otimes \underline{a}$, we reinitialize at $b_0 a_1$, the 1-st component of $\underline{b} \otimes \underline{a}$ then stride through $\underline{b} \otimes \underline{a}$ with length the size of $\underline{a}$. This permutation of data continues until we form $\underline{a} \otimes \underline{b}$. This procedure is an example of the important notion of a stride permutation. Stride permutation will be discussed in great detail beginning in the next section.

The tensor product of an $R \times S$ matrix A with an $M \times N$ matrix B is the $RM \times SN$ matrix, $A \otimes B$, given by

$$\begin{bmatrix} a_{00} B & a_{01} B & \cdots & a_{0,S-1} B \\ \cdot & & & \\ \cdot & & & \\ \cdot & & & \\ a_{R-1,0} B & & \cdots & a_{R-1,S-1} B \end{bmatrix}. \tag{17}$$

Setting $C = A \otimes B$, the coefficients of C are given by

$$c_{m+rM,n+sN} = a_{r,s}\, b_{m,n}. \tag{18}$$

It is natural to view the tensor product $A \otimes B$ as being formed from blocks of scalar multiples of B. The relationship between tensor products of matrices and vectors is contained in the next result.

Theorem 1. If A is an $R \times S$ matrix and B is an $M \times N$ matrix, then

$$(A \otimes B)(\underline{a} \otimes \underline{b}) = A\underline{a} \otimes B\underline{b}, \qquad (19)$$

for any vectors $\underline{a}$ and $\underline{b}$ of sizes S and N, respectively.

Proof The vector $\underline{a} \otimes \underline{b}$

$$\begin{bmatrix} a_0\underline{b} \\ a_1\underline{b} \\ . \\ . \\ . \\ a_{s-1}\underline{b} \end{bmatrix}$$

can be viewed as consisting of consecutive segments

$$a_0\underline{b} \, , \, a_1\underline{b} \, , \, \cdots \, , \, a_{s-1}\underline{b} \qquad (20)$$

each of size N. The first M rows of $(A \otimes B)(\underline{a} \otimes \underline{b})$ is

$$\left(a_{0,0}a_0 + a_{0,1}a_1 + \cdots + a_{0,s-1}a_{s-1}\right) B \, \underline{b}. \qquad (21)$$

Continuing in this way proves the theorem.

Denote by $\mathbb{C}^L$ the vector space of all complex vectors of size L. As $\underline{a}$ runs over all vectors of size L and $\underline{b}$ runs over all vectors of size M, the tensor products $\underline{a} \otimes \underline{b}$ span the space $\mathbb{C}^{LM}$ (see problems 4 and 5). To prove that the action of two matrix expressions on $\mathbb{C}^{LM}$ are equal, we only need to prove that they are equal on the $\underline{a} \otimes \underline{b}$ with $\underline{a} \in \mathbb{C}^L$ and $\underline{b} \in \mathbb{C}^M$.

Theorem 2. If A and C are $L \times L$ matrices and B and D are $M \times M$ matrices, then

$$(A \otimes B)(C \otimes D) = AC \otimes BD. \qquad (22)$$

Proof Take vectors $\underline{a}$ and $\underline{b}$ of sizes L and M respectively. By (20),

$$(A \otimes B)(C \otimes D)(\underline{a} \otimes \underline{b}) = (A \otimes B)(C\underline{a} \otimes D\underline{b})$$

$$= AC\underline{a} \otimes BD\underline{b},$$

proving (22), in light of the proceeding discussion.

More generally, the tensor products

$$\underline{a} \otimes \underline{b} \otimes \underline{c} = \underline{a} \otimes (\underline{b} \otimes \underline{c}) = (\underline{a} \otimes \underline{b}) \otimes \underline{c}, \quad \underline{a} \in \mathbb{C}^L, \, \underline{b} \in \mathbb{C}^M, \, \underline{c} \in \mathbb{C}^N, \tag{23}$$

span the space $\mathbb{C}^{LMN}$ and the observation about matrix expressions can be extended.

An important special case of formula (22) is the following decomposition. Denote by I_L the $L \times L$ identity matrix. Then

$$A \otimes B = (I_M \otimes B)(A \otimes I_L) = (A \otimes I_L)(I_M \otimes B), \tag{25}$$

where A is an $M \times M$ matrix and B is an $L \times L$ matrix. In order to better understand the computation $(A \otimes B)\underline{x}$, we need to examine the factors $I_M \otimes B$ and $A \otimes I_L$.

$I_M \otimes B$ is the direct sum of M copies of B,

$$I_M \otimes B = \begin{bmatrix} B & & & \\ & \cdot & & 0 \\ & & \cdot & \\ 0 & & \cdot & \\ & & & B \end{bmatrix} \tag{26}$$

and its action on $\underline{x}$ is the action of B on the M consecutive segments of $\underline{x}$ of size L. This direct sum can be computed in *parallel*, and we call $I_M \otimes B$ a *parallel operation*.

$$\underline{x} \quad \rightarrow \quad \begin{matrix} X_0 \\ X_1 \\ \cdot \\ \cdot \\ \cdot \\ X_{M-1} \end{matrix} \quad \rightarrow \quad \begin{matrix} B \\ B \\ \cdot \\ \\ \\ B \end{matrix} \quad \rightarrow \quad \begin{matrix} BX_0 \\ BX_1 \\ \cdot \\ \cdot \\ \cdot \\ BX_{M-1} \end{matrix} \quad (I_M \otimes B)X_2$$

To the vector $\underline{x}$ of size LM, associate the $L \times M$ matrix

$$X = [X_0 \ \ X_1 \ \ \cdots \ \ X_{M-1}]. \tag{27}$$

The vector $(I_M \otimes B)\underline{x}$ corresponds to the matrix

$$BX = [\, BX_0 \quad BX_1 \quad \cdots \quad BX_{M-1} \,]. \tag{28}$$

The computation $(A \otimes I_L)\, \underline{x}$ can be interpreted as a *vector operation* of A on the vectors $X_0, X_1, \cdots, X_{M-1}$. In fact,

$$(A \otimes I_L)\, \underline{x} = \begin{bmatrix} a_{0,0}X_0 + \cdots + a_{0,M-1}X_{M-1} \\ \vdots \\ a_{M-1,0}X_0 + \cdots + a_{M-1,M-1}X_{M-1} \end{bmatrix}. \tag{29}$$

Operations involved in (29) are scalar-vector multiplication and vector addition. Note that the vector in (29) corresponds to the matrix XA^t. Hence $(A \otimes B)\underline{x}$ corresponds to BXA^t.

Factorization (25) decomposes the computation of $(A \otimes B)\underline{x}$ into the parallel operation $I_M \otimes B$ followed by the vector operation $A \otimes I_L$.

2.3. Stride Permutations

In this section, we discuss the it stride permutations that govern the data flow required to parallelize or vectorize a tensor product computation and as we will see in the next chapter, play a crucial role in the implementation of FT computations. On some machines the action of a stride permutation can be implemented as elements of the input vector are loaded from main memory into registers. For architectures where this is the case, considerable savings can be obtained by performing these permutations when loading the input into register. Take $N = 2M$. The tensor products $\underline{a} \otimes \underline{b}$ where $\underline{a} \in \mathbb{C}^2$ and $\underline{b} \in \mathbb{C}^M$ span $\mathbb{C}^N$. We define the N-point stride M permutation matrix $P(N, M)$ by the rule

$$P(N,M)(\underline{a} \otimes \underline{b}) = \underline{b} \otimes \underline{a}, \quad \underline{a} \in \mathbb{C}^2, \quad \underline{b} \in \mathbb{C}^M. \tag{1}$$

If $\underline{x}$ is a vector of size N then computation $P(N, M)\underline{x}$ can be given as follows. Associate to $\underline{x}$ the $M \times 2$ array

$$X = \begin{bmatrix} x_0 & x_M \\ x_1 & x_{M+1} \\ \vdots & \vdots \\ x_{M-1} & x_{N-1} \end{bmatrix}. \tag{2}$$

Then

$$X^t = \begin{bmatrix} x_0 & x_1 & \cdots & x_{M-1} \\ x_M & x_{M+1} & \cdots & x_{N-1} \end{bmatrix}, \tag{3}$$

and we have

$$P(N, M)\underline{x} = \begin{bmatrix} y_0 \\ \vdots \\ y_{m-1} \end{bmatrix}, \quad \text{where} \quad y_r = \begin{bmatrix} x_r \\ x_r + m \end{bmatrix} \tag{4}$$

The matrix $P(N, M)$ is usually called the *perfect shuffle*. It strides through $\underline{x}$ with stride of length M.

Example 1. Take $N = 4$. Then

$$P(4, 2) = \begin{bmatrix} 1 & 0 & 0 & 0 \\ 0 & 0 & 1 & 0 \\ 0 & 1 & 0 & 0 \\ 0 & 0 & 0 & 1 \end{bmatrix}, \quad P(4, 2)\underline{x} = \begin{bmatrix} x_0 \\ x_2 \\ x_1 \\ x_3 \end{bmatrix}.$$

Example 2. Take $N = 8$. Then

$$P(8, 4) = \begin{bmatrix} 1 & 0 & 0 & 0 & 0 & 0 & 0 & 0 \\ 0 & 0 & 0 & 0 & 1 & 0 & 0 & 0 \\ 0 & 1 & 0 & 0 & 0 & 0 & 0 & 0 \\ 0 & 0 & 0 & 0 & 0 & 1 & 0 & 0 \\ 0 & 0 & 1 & 0 & 0 & 0 & 0 & 0 \\ 0 & 0 & 0 & 0 & 0 & 0 & 1 & 0 \\ 0 & 0 & 0 & 1 & 0 & 0 & 0 & 0 \\ 0 & 0 & 0 & 0 & 0 & 0 & 0 & 1 \end{bmatrix},$$

and

$$P(8, 4)\underline{x} = \begin{bmatrix} x_0 \\ x_4 \\ x_1 \\ x_5 \\ x_2 \\ x_6 \\ x_3 \\ x_7 \end{bmatrix}.$$

Example 3. Take $N = 6$. Then

$$P(6,3) = \begin{bmatrix} 1 & 0 & 0 & 0 & 0 & 0 \\ 0 & 0 & 0 & 1 & 0 & 0 \\ 0 & 1 & 0 & 0 & 0 & 0 \\ 0 & 0 & 0 & 0 & 1 & 0 \\ 0 & 0 & 1 & 0 & 0 & 0 \\ 0 & 0 & 0 & 0 & 0 & 1 \end{bmatrix},$$

and

$$P(6,3)\underline{x} = \begin{bmatrix} x_0 \\ x_3 \\ x_1 \\ x_4 \\ x_2 \\ x_5 \end{bmatrix}.$$

Suppose now that $N = RS$. The *N-point stride S permutation matrix* $P(N,S)$ is defined by

$$P(N,S) = (\underline{a} \otimes \underline{b}) = \underline{b} \otimes \underline{a}, \tag{5}$$

where $\underline{a}$ and $\underline{b}$ are arbitrary vectors of sizes R and S, respectively. The action of $P(N,S)$ on an arbitrary vector $\underline{x}$ of size N can be described as follows. To $\underline{x}$, associate the $S \times R$ matrix

$$X = \begin{bmatrix} x_0 & x_S & \cdots & x_{(R-1)S} \\ x_1 & & & \\ \vdots & & & \\ x_{S-1} & x_{2S-1} & \cdots & x_{N-1} \end{bmatrix}. \tag{6}$$

Then $P(N,S)\underline{x}$ corresponds to the transpose of X. Writing

$$X^t = [Y_0 \quad Y_1 \quad \cdots \quad Y_{S-1}], \tag{7}$$

where

$$Y_k = \begin{bmatrix} x_k \\ x_{k+S} \\ \vdots \\ x_{k+(R-1)S} \end{bmatrix}, \tag{8}$$

we have

$$P(N,S)\underline{x} = \begin{bmatrix} Y_0 \\ Y_1 \\ \vdots \\ Y_{S-1} \end{bmatrix}. \tag{9}$$

To compute $P(N,S)\underline{x}$, we stride through $\underline{x}$ with stride S. Formula (1) will be used repeatedly to derive matrix identities involving stride permutations.

The algebra of stride permutation has an important impact on the design of tensor product algorithms. We begin this study with the next result.

Theorem 1. If $N = RST$ then

$$P(N,ST) = P(N,S)P(N,T). \tag{10}$$

proof Take vectors $\underline{a} \in \mathbb{C}^R, \underline{b} \in \mathbb{C}^S$ and $\underline{c} \in \mathbb{C}^T$. Then

$$P(N,ST)(\underline{a} \otimes \underline{b} \otimes \underline{c}) = \underline{b} \otimes \underline{c} \otimes \underline{a},$$

and

$$P(N,S)P(N,T)(\underline{a} \otimes \underline{b} \otimes \underline{c}) = P(N,S)(\underline{c} \otimes \underline{a} \otimes \underline{b}) = \underline{b} \otimes \underline{c} \otimes \underline{a},$$

proving the theorem.

In particular, from theorem 1,

$$P(NM,M)^{-1} = P(NM,N). \tag{11}$$

Example 4. Take $N = 4 \times 2$. Then

$$P(8,2)\underline{x} = \begin{bmatrix} x_0 \\ x_2 \\ x_4 \\ x_6 \\ x_1 \\ x_3 \\ x_5 \\ x_7 \end{bmatrix}, \quad P(8,2)^2\underline{x} = \begin{bmatrix} x_0 \\ x_4 \\ x_1 \\ x_5 \\ x_2 \\ x_6 \\ x_3 \\ x_7 \end{bmatrix},$$

from which we see that

$$P(8,2)^2 = P(8,4),$$

$$P(8,2)^3 = I_8.$$

In general, an $N \times N$ permutation matrix can be given by a permutation of Z/N. Let π be a permutation of Z/N. We represent π using the following notation.

$$\pi = (\pi(0), \pi(1), \ldots, \pi(N-1)). \qquad (12)$$

Define the $N \times N$ permutation matrix $P(\pi)$ by the condition

$$P(\pi)\underline{x} = \underline{y} \qquad (13)$$

where

$$y_j = x_{\pi(j)}, \quad 0 \le j < N. \qquad (14)$$

Example 5. Take $N = 8$ and

$$\pi = (0,4,1,5,2,6,3,7).$$

Then

$$P(\pi)\underline{x} = \begin{bmatrix} x_0 \\ x_4 \\ x_1 \\ x_5 \\ x_2 \\ x_6 \\ x_3 \\ x_7 \end{bmatrix}$$

and we see that

$$P(\pi) = P(8,4).$$

Example 6. Take $N = 12$ and

$$\pi = (0,3,6,9,1,4,7,10,2,5,8,11).$$

Then

$$P(\pi) = P(12, 3).$$

Direct computation shows that the mapping

$$\pi \longrightarrow P(\pi)^{-1} \qquad (15)$$

is a group-isomorphism from the group of permutations of Z/N under composition onto the group of $N \times N$ permutation matrices under matrix product. In fact

$$P(\pi_2 \cdot \pi_1) = P(\pi_1)P(\pi_2), \qquad (16)$$

$$P(\pi^{-1}) = P(\pi)^{-1}. \qquad (17)$$

Suppose $N = RS$. Take any n, $0 \le n < N$. Then n can be written uniquely as

$$n = a + bS, \quad 0 \le a < S, \quad 0 \le b < R, \qquad (18)$$

as well as

$$n = c + dR, \quad 0 \le c < R, 0 \le d < S. \qquad (19)$$

Corresponding to the factorization $N = RS$, we define the permutation π of Z/N by the setting

$$\pi(a + bS) = b + aR, \quad 0 \le a < S, \quad 0 \le b < R. \qquad (20)$$

Example 7. Take $N = 8$, $R = 2$ and $S = 4$. Then

$$\pi = (0, 2, 4, 6, 1, 3, 5, 7)$$

and

$$P(\pi) = P(8, 2).$$

Example 8. Take $N = 12$, $S = 3$ and $T = 4$. Then

$$\pi = (0, 3, 6, 9, 1, 4, 7, 10, 2, 5, 8, 11)$$

and
$$P(\pi) \; = \; P(12,3).$$

In general, if $N = RS$ and π is the permutation of Z/N defined by (20) then,
$$P(\pi) \; = \; P(N,S). \tag{21}$$

Consider the set of $N \times N$ permutation matrices
$$\{P(N,S) \mid S/N\}. \tag{22}$$

We will describe the permutation matrices in this set in terms of the unit group $U(N-1)$ of $Z/(N-1)$. The unit group $U(N-1)$ is given by
$$U(N-1) \; = \; \{0 \le T < N-1 \mid (T, N-1) = 1\}. \tag{23}$$

If $T \in U(N-1)$ then multiplication by $T \bmod (N-1)$ is a bijection of the set
$$\{0, 1, \ldots, N-2\}. \tag{24}$$

Define the permutation π_T of Z/N by the two rules
$$\pi_T(k) \equiv kT \bmod (N-1), \quad 0 \le k < N-1, \tag{25}$$

$$\pi_T(N-1) \; = \; (N-1) \tag{25}$$

Observe that if T/N then $(T, N-1) = 1$ and we can define π_T.

Example 9. Take $N = 12, S = 3$ and $T = 4$. Then

$$\pi_4 \; = \; (0,4,8,1,5,9,2,6,10,3,7,11).$$

We see that

$$\pi_4(a+3b) \; = \; b + 4a, \quad 0 \le a < 3, \; 0 \le b < 4,$$

and

$$P(\pi_4) \; = \; P(12,4).$$

Theorem 2. If $N = RS$ then

$$P(\pi_S) = P(N, S) \tag{27}$$

Proof We will show that

$$\pi_S(a + bR) = b + aS, \quad 0 \le a < RS, \quad 0 \le b < S. \tag{28}$$

$$N - 1 = RS - 1 \equiv 0 \; mod \; (N - 1). \tag{29}$$

$$\pi_S(a + bR) \equiv aS + bRS$$

$$\equiv aS + b + b(RS - 1) \equiv aS + b \; mod \; (N - 1). \tag{30}$$

Consider the set of $N \times N$ permutation matrices

$$\{P(\pi_S) \mid S \in U(N - 1)\}. \tag{31}$$

This set is a group isomorphic to $U(N - 1)$. In fact

$$P(\pi_R)P(\pi_S) = P(\pi_U), \quad U \equiv RS \; mod \; (N - 1) \tag{32}$$

$$P(\pi_R)^{-1} = P(\pi_{R^{-1}}), \quad R^{-1} \; taken \; mod \; (N - 1). \tag{33}$$

Theorem 3. If $N = 2^M$, the set

$$\{P(2^M, 2^m) \mid 0 \le m < M\} \tag{34}$$

is a cyclic group generated by $P(2^M, 2)$. In fact,

$$P(2^M, 2^m)P(2^M, 2^k) = P(2^M, 2^{m+k}) \tag{35}$$

where $m + k$ is taken $mod \; M$.

Proof Consider integers $0 \le m, k < M$. If $m + k < M$ then there is nothing to prove. Suppose that $M \le m + k$. Set $l = m + k - M$. We have $0 \le l < M$ and $l \equiv m + k \; mod \; M$.

$$2^{m+k} = 2^{l+M}$$

$$2^{m+k} - 2^l = 2^{l+M} - 2^l = 2^l(2^M - 1) \equiv 0 \; mod \, (2^M - 1)$$

It follows that

$$2^l \equiv 2^m 2^k \; mod \, (2^M - 1),$$

proving the theorem.

More generally, we have the following result which we state without proof.

Theorem 4. If p is a prime, then the set

$$\{P(p^R, p^S) \mid S/R\} \tag{36}$$

is a cyclic group of order R generated by $P(p^R, p)$.

It is sometimes useful to represent permutations and general computations by diagrams which give a picture of data flow.

Example 10. The permutation $P(4, 2)$ can be represented by

$$
\begin{array}{ll}
x_0 \cdot & \cdot x_0 \\
x_1 \cdot & \cdot x_2 \\
x_2 \cdot & \cdot x_1 \\
x_3 \cdot & \cdot x_3
\end{array}
$$

Example 11. The permutation $I_2 \otimes P(4, 2)$ and $P(4, 2) \otimes I_2$ can be represented by

$$
\begin{array}{llll}
x_0 \cdot & \cdot x_0 & x_0 \cdot & \cdot x_0 \\
x_1 \cdot & \cdot x_2 & x_1 \cdot & \cdot x_1 \\
x_2 \cdot & \cdot x_1 & x_2 \cdot & \cdot x_4 \\
x_3 \cdot & \cdot x_3 & x_3 \cdot & \cdot x_5 \\
x_4 \cdot & \cdot x_4 & x_4 \cdot & \cdot x_2 \\
x_5 \cdot & \cdot x_6 & x_5 \cdot & \cdot x_3 \\
x_6 \cdot & \cdot x_5 & x_5 \cdot & \cdot x_6 \\
x_7 \cdot & \cdot x_7 & x_7 \cdot & \cdot x_7
\end{array}
$$

Example 12. The permutations $P(8,2)$ and $P(8,4)$ can be represented by

$x_0\cdot$	$\cdot x_0$	$x_0\cdot$	$\cdot x_0$
$x_1\cdot$	$\cdot x_2$	$x_1\cdot$	$\cdot x_4$
$x_2\cdot$	$\cdot x_4$	$x_2\cdot$	$\cdot x_1$
$x_3\cdot$	$\cdot x_6$	$x_3\cdot$	$\cdot x_5$
$x_4\cdot$	$\cdot x_1$	$x_4\cdot$	$\cdot x_2$
$x_5\cdot$	$\cdot x_3$	$x_5\cdot$	$\cdot x_6$
$x_6\cdot$	$\cdot x_5$	$x_6\cdot$	$\cdot x_3$
$x_7\cdot$	$\cdot x_7$	$x_7\cdot$	$\cdot x_7$

We see from the diagram of Example 11 that $I_2 \otimes P(4,2)$ consists of two parallel copies of $P(4,2)$. To compute the action of $P(4,2) \otimes I_2$, we can first form the vectors

$$\underline{x}(0) = \begin{bmatrix} x_0 \\ x_1 \end{bmatrix}, \ \underline{x}(1) = \begin{bmatrix} x_2 \\ x_3 \end{bmatrix}, \ \underline{x}(2) = \begin{bmatrix} x_4 \\ x_5 \end{bmatrix}, \ \underline{x}(3) = \begin{bmatrix} x_6 \\ x_7 \end{bmatrix}. \qquad (37)$$

and compute the vector operation of $P(4,2)$ on these four vectors,

$$P(4,2) \begin{bmatrix} \underline{x}(0) \\ \underline{x}(1) \\ \underline{x}(2) \\ \underline{x}(3) \end{bmatrix}. \qquad (38)$$

In the preceeding section, we discussed how certain tensor product expressions could be viewed as vector operations, parallel operations, or as a combination of vector and parallel operations. An important tool for interchanging the operations in a given algorithm is the *commutation theorem*.

Theorem 5. If A is an $R \times R$ matrix and B is an $S \times S$ matrix, then

$$P(A \otimes B)P^{-1} = B \otimes A, \qquad (40)$$

where $P = P(N,S)$ and $N = RS$.

Proof Set $\underline{z} = \underline{x} \otimes \underline{y}$ where $\underline{x}$ is and R-dimensional vector and $\underline{y}$ is an S-dimensional vector. Then by definition,

$$(A \otimes B)(\underline{x} \otimes \underline{y}) = A\underline{x} \otimes B\underline{y}, \qquad (41)$$

$$P(A \otimes B)\underline{z} = B\underline{y} \otimes A\underline{x}. \tag{42}$$

Arguing in the same way,

$$(B \otimes A)P\underline{z} = B\underline{y} \otimes A\underline{x}, \tag{43}$$

proving the theorem.

Corollary

$$P(I_R \otimes B)P^{-1} = B \otimes I_R, \tag{44}$$

where $P = P(N, S)$ and $N = RS$. As an important application of the commutation theorem, we observe that

$$A \otimes B = (A \otimes I_S)P(N, R)(B \otimes I_R)P(N, R)^{-1}, \tag{45}$$

$$A \otimes B = P(N, R)(I_S \otimes A)P(N, R)^{-1}(I_R \otimes B). \tag{46}$$

Factorization (45) decomposes $A \otimes B$ into a sequence of vector operations; the first operates on vectors of size R while the second operates on vectors of size S. The intervening stride permutations provide a mathematical language for describing the readdressing between stages of the computation. In the same way, we interpret (46) as a sequence of parallel operations.

2.4. Multidimensional Tensor Products

Tensor product identities will be used to obtain important factorizatons of multidimensional tensor products, which can then be applied to implementation problems. The rules of implementation established in this section will have important consequences in the rest of this text. The first application will be to the various Cooley-Tukey like FFT algorithms in the next chapter. The stride permutations appearing in these factorizations will make explicit the readdressing needed to carryout computations. We begin with an example. Take positive integers N_1, N_2 and N_3. Set $N = N_1 N_2 N_3$. A_N denotes any $N \times N$ matrix. The product rule implies

$$A_{N_1} \otimes A_{N_2} \otimes A_{N_3}$$

$$= (A_{N_1} \otimes I_{N_2 N_3})(I_{N_1} \otimes A_{N_2} \otimes I_{N_3})(I_{N_1 N_2} \otimes A_{N_3}) \qquad (1)$$

The factor $A_{N_1} \otimes I_{N_2 N_3}$ is a vector operation while the factor $I_{N_1 N_2} \otimes A_{N_3}$ is a parallel operation. The middle factor, $I_{N_1} \otimes A_{N_2} \otimes I_{N_3}$, is of mixed type involving N_1 copies of the vector operation $A_{N_2} \otimes I_{N_3}$. There are several ways of modifying these factors, using the commutation theorem. We can parallelize (1) by the formulas

$$A_{N_1} \otimes I_{N_2 N_3} = P_1(I_{N_2 N_3} \otimes A_{N_1})P_1^{-1}, \qquad (2)$$

$$I_{N_1} \otimes A_{N_2} \otimes I_{N_3} = P_2(I_{N_2 N_3} \otimes A_{N_2})P_2^{-1}, \qquad (3)$$

where $P_1 = P(N, N_1)$ and $P_2 = P(N, N_1 N_2)$. We can rewrite (1) as

$$A_{N_1} \otimes A_{N_2} \otimes A_{N_3}$$

$$= P_1(I_{N_2 N_3} \otimes A_{N_1})P_1^{-1}P_2(I_{N_1 N_3} \otimes A_{N_2})P_2^{-1}(I_{N_1 N_2} \otimes A_{N_3}). \quad (4)$$

A second parallelization of (1) comes from replacing (3) by

$$I_{N_1} \otimes A_{N_2} \otimes I_{N_3} = Q_2(I_{N_1 N_3} \otimes A_{N_2})Q_2^{-1}, \qquad (5)$$

where $Q_2 = I_{N_1} \otimes P(N_2 N_3, N_2)$. This leads to the factorization

$$A_{N_1} \otimes A_{N_2} \otimes A_{N_3}$$

$$= P_1(I_{N_2 N_3} \otimes A_{N_1})P_1^{-1}Q_2(I_{N_1 N_3} \otimes A_{N_2})Q_2^{-1}(I_{N_1 N_2} \otimes A_{N_3}), \quad (6)$$

These two parallel factorizations differ only in data flow. In the first the readdressing between the computational stages is given by P_2^{-1}, $P_1^{-1}P_2$ and P_1 while in the second the readdressing is given by Q_2^{-1}, $P_1^{-1}Q_2$ and P_1. Each will have advantages and disadvantages which can be made explicit when implementing on a specific computer.

In general, the permutations that arise from commuting terms in a multidimensional tensor product are built up from products of terms of the form $I \otimes P \otimes I$, where I denotes an identity matrix and P denotes a stride permutation. In particular, $I_{N_1} \otimes P(N_2 N_3, N_3)$ is N_1 copies of the permutation $P(N_2 N_3, N_3)$. As such, it performs a stride permutation on N_1 segments of the input vector beginning at different offsets. It can be implemented as a loop of stride permutations, where the same permutation is performed, but the initial offset is incremented by $N_2 N_3$ at each iteration. The second type of permutation can be thought of as permuting blocks of the input vector. Thus $P(N_2 N_3, N_3) \otimes I_{N_1}$ permutes segments of length N_1 at stride N_3. This can be implemented by loading blocks of N_1 consecutive elements, beginning at offsets given by the permutation $P(N_2 N_3, N_3)$.

Factorizations of the permutation occuring in computing terms in multidimensional tensor products offer programming options that can be used to match the algorithm computing the action of these multidimensional tensor products to specific machine architecture. Depending on machine parameters such as maximal vector length, minimal vector length, number of processors and communication network a full parallelization or vectorization may not be desired. The rules established can be modified to partially parallelize or vectorize. The next result describes a factorization which is especially useful

Theorem 1. If $N = N_1 N_2 N_3$ then

$$P(N, N_3) = (P(N_1 N_3, N_3) \otimes I_{N_2})(I_{N_1} \otimes P(N_2 N_3, N_3)). \quad (7)$$

Proof Take $\underline{a} \in \mathbb{C}^{N_1}$, $\underline{b} \in \mathbb{C}^{N_2}$ and $\underline{c} \in \mathbb{C}^{N_3}$. Then

$$(P(N_1 N_3, N_3) \otimes I_{N_2})(I_{N_2} \otimes P(N_2 N_3, N_3))(\underline{a} \otimes \underline{b} \otimes \underline{c})$$

$$= (P(N_1 N_3, N_3) \otimes I_{N_2})(\underline{a} \otimes \underline{c} \otimes \underline{b})$$

$$= \underline{c} \otimes \underline{a} \otimes \underline{b}$$

$$= P(N, N_3)(\underline{a} \otimes \underline{b} \otimes \underline{c}),$$

proving the theorem.

Suppose now positive integers $N_1, N_2, \cdots, N_t$ are taken. Denote an arbitrary $N_k \times N_k$ matrix by A_{N_k}. Set $N = N_1 \cdots N_t$, $N(k) = N_1 N_2 \cdots N_k$ and $N(0) = 1$. The product rule implies

$$A_{N_1} \otimes \cdots \otimes A_{N_t} = \prod_{k=1}^{t} I_{N(k-1)} \otimes A_{N_k} \otimes I_{N/N(k)}. \tag{8}$$

Set $P_k = P(N, N(k))$. Then, by the commutation theorem,

$$I_{N(k-1)} \otimes A_{N_k} \otimes I_{N/N_k} = P_k(I_{N/N_k} \otimes A_{N_k})P_k^{-1}. \tag{9}$$

and we can parallelize (8) by the factorization given in the next theorem.

Theorem 2. If $P_k = P(N, N(k))$ then

$$A_{N_1} \otimes \cdots \otimes A_{N_t} = \prod_{k=1}^{t} P_k(I_{N/N_k} \otimes A_{N_k})P_k^{-1}. \tag{10}$$

Compare (10) and (4). The description of the intervening permutation can be simplified by combining permutations. For example in (4)

$$P_1^{-1}P_2 = P(N, N_1 N_2)P(N, N_1 N_2) = P(N, N_2), \quad N = N_1 N_2 N_3. \tag{11}$$

and we can rewrite (4) as

$$A_{N_1} \otimes A_{N_2} \otimes A_{N_3} = P(N, N_1)(I_{N_2 N_3} \otimes A_{N_1})$$

$$P(N, N_2)(I_{N_1 N_3} \otimes A_{N_2})P(N, N_3)(I_{N_1 N_2} \otimes A_{N_3}). \tag{12}$$

If $M = N_1 = N_2 = N_3$ and A, B and C are $M \times M$ matrices, factorization (12) becomes

$$A \otimes B \otimes C = P(I_{M^2} \otimes A)P(I_{M^2} \otimes B)P(I_{M^2} \otimes C), \tag{13}$$

where $P = P(M^3, M)$. In this case, the readdressing between each of the stages of the computation is the same and is given by P. Using

$$P_k^{-1} P_{k+1} = P(N, N/N(k)) P(N, N(k+1)) = P(N, N_{k+1}) \quad (14)$$

in theorem 2, we have the next result.

Theorem 3.

$$A_{N_1} \otimes \cdots \otimes A_{N_t} = \prod_{k=1}^{t} P(N, N_k)(I_{N/N_k} \otimes A_{N_k}).$$

As in the example, if $N_1 = N_2 = \cdots = N_t$ then the readdressing between computational stages is exactly the same.

A second parallelization of (8) can be obtained from the identity

$$I_{N(k-1)} \otimes A_{N_k} \otimes I_{N/N(k)} = Q_k (I_{N/N_k} \otimes A_{N_k}) Q_k^{-1}, \quad (15)$$

where $Q_k = I_{N(k-1)} \otimes P(N/N(k-1), N_k)$. This leads to the next result.

Theorem 4.

$$A_{N_1} \otimes \ldots \otimes A_{N_t} = \prod_{k=1}^{t} Q_k (I_{N/N_k} \otimes A_{N_k}) Q_k^{-1},$$

where

$$Q_k = I_{N(k-1)} \otimes P(N/N(k-1), N_k).$$

Vector factorization arises in the same way. For instance

$$I_{N/N_k} \otimes A_{N_k} = P(N, N/N_k)(A_{N_k} \otimes I_N/N_k) P(N, N/N_k)^{-1} \quad (16)$$

and we can obtain the vector factorization analog of theorem 3.

Theorem 5.

$$A_{N_1} \otimes \ldots \otimes A_{N_t} = \prod_{k=1}^{t} (A_{N_k} \otimes I_{N/N_k}) P(N, N_k).$$

2.5. Vector Implementation of Tensor Products

In this section, tensor product identities will be used to design algorithms computing tensor product operations on a sample vector processor. Our model of a vector processor includes main memory, vector registers and a communication network between main memory and vector registers which will be described in detail as required. Vector operations are performed on vectors located in vector registers. Some standard vector operations are vector addition, subtraction and scalar-vector and vector multiplication. To take advantage of the high speed computational rate offered by vector operation, it is essential to keep memory transfers to a minimum and to perform vector operations on vectors residing in vector registers as much as possible. Also, the transfer of data between main memory and vector registers, on many processors, is especially suited to implement the stride permutations arising from tensor product operations.

There are several key machine parameters which must be kept in mind when designing algorithms. First vector registers have a maximum size which limits the size of vectors that can be used on vector instructions. Also, due to 'start up costs', there is usually a lower bound on the size of vectors which can efficiently be operated on by vector operations. If a computation requires operations on larger vectors than allowed, then the computation must be segmented and several vector instructions combined to perform the computation. The language of tensor products is ideally suited to design algorithms which satisfy this key design parameter.

Memory transfer can also be performed with vector operations. These vector operations correspond to stride permutations. A vector of elements in main memory can be *loaded* into a vector register with the following instruction.

$$\bullet VI \quad X, \quad S$$

The vector of elements in memory having initial address X is loaded into the vector register VI at stride S. A special register called the vector length register VL determines the number of elements which are loaded. For example, if

$$X = \begin{bmatrix} x_0 \\ x_1 \\ x_2 \\ x_3 \\ x_4 \\ x_5 \end{bmatrix}, \quad VL = 3,$$

then

$$\bullet V0 \quad X, \quad 2$$

loads the vector register $V0$ with the elements of X beginning at x_0 with stride 2,

$$V0 = \begin{bmatrix} x_0 \\ x_2 \\ x_4 \end{bmatrix}.$$

The result of the load instruction

$$\bullet V1 \quad X+1, \quad 2$$

is

$$V1 = \begin{bmatrix} x_1 \\ x_3 \\ x_5 \end{bmatrix}.$$

The memory transfer operation which takes a vector in memory of size MN and fills at stride N, N vector registers with vectors of size M will be denoted by L_N^{MN}. Thus

$$L_2^6 \sim P(6,2) \qquad \begin{bmatrix} x_0 \\ x_1 \\ x_2 \\ x_3 \\ x_4 \\ x_5 \end{bmatrix} \xrightarrow{L_2^6} \begin{bmatrix} x_0 \\ x_2 \\ x_4 \end{bmatrix} \quad \begin{bmatrix} x_1 \\ x_3 \\ x_5 \end{bmatrix}.$$

$$\text{mainmemory} \qquad\qquad V0 \qquad V1$$

The contents of a vector register can be *stored* into main memory with the instruction

$$\bullet, Y, \quad S \quad VK.$$

The contents of the vector register VK are stored in memory having initial address Y at stride S. For example, if

$$V0 = \begin{bmatrix} x_0 \\ x_1 \end{bmatrix},$$

then the result of the vector instruction

$$\bullet, Y, \quad 3 \quad V0$$

is

$$Y = x_0 \quad -- \quad -- \quad x_1 \quad --- \quad ---$$

If

$$V1 = \begin{bmatrix} x_2 \\ x_3 \end{bmatrix}, V0 = \begin{bmatrix} x_4 \\ x_5 \end{bmatrix},$$

then the result of the sequence of store instructions,

$$\bullet, Y, \quad 3 \quad V0$$

$$\bullet, Y+1, \quad 3 \quad V2$$

$$\bullet, Y+2, \quad 3 \quad V3$$

is the sequence of stores

$$Y = x_0 \quad -- \quad -- \quad x_1 \quad --- \quad ---$$

$$Y = x_0 \quad x_2 \quad -- \quad x_1 \quad x_3 \quad ---$$

$$Y = x_0 \quad x_2 \quad x_4 \quad x_1 \quad x_3 \quad x_5.$$

The memory transfer operation which takes the contents of N vector registers with vector size M and stores them in memory with stride N will be denoted by S_N^{MN}.

$$S_N^{MN} L_N^{MN} = I.$$

$$(L_3^6)^{-1} = S_3^6 = P(6,2) \quad \begin{bmatrix} x_0 \\ x_1 \end{bmatrix} \quad \begin{bmatrix} x_2 \\ x_3 \end{bmatrix} \quad \begin{bmatrix} x_4 \\ x_5 \end{bmatrix} \quad \overset{S_3^6}{\to} \quad \begin{bmatrix} x_0 \\ x_2 \\ x_4 \\ x_1 \\ x_3 \\ x_5 \end{bmatrix} .$$

$$\qquad\qquad\qquad\quad V0 \qquad V1 \qquad V2 \qquad\quad \text{memory}$$

A load-stride followed by a store-stride can carry out a stride permutation. The stride permutation $P(6,2)$ can be implemented with a sequence of operations. Take $VL = 2$.

$$\bullet V0 \quad X, \quad 1$$

$$\bullet V1 \quad X+2, \quad 1 \quad \text{load at stride 1}$$

$$\bullet V2 \quad X+4, \quad 1$$

$$\bullet, Y, \quad 3 \quad V0$$

$$\bullet, Y+1, \quad 3 \quad V1 \quad \text{store at stride 3}$$

$$\bullet, Y+2, \quad 3 \quad V2$$

Tensor product operations of the form $A \otimes I_N$ can be implemented directly with vector instructions as long as N is less than equal to the maximum vector register length. For example, if

$$y = (A \otimes I_3)x,$$

where

$$A = \begin{bmatrix} 1 & 1 \\ 1 & -1 \end{bmatrix},$$

then

$$y = \begin{bmatrix} x_0 + x_3 \\ x_1 + x_4 \\ x_2 + x_5 \\ x_0 - x_3 \\ x_1 - x_4 \\ x_2 - x_5 \end{bmatrix}.$$

If

$$VO = \begin{bmatrix} x_0 \\ x_1 \\ x_2 \end{bmatrix}, \quad V1 = \begin{bmatrix} x_3 \\ x_4 \\ x_5 \end{bmatrix},$$

then y is computed by the vector instructions

$$\bullet V2 \quad V0 + V1$$

$$\bullet V2 \quad V0 - V1.$$

The first instruction is the vector addition of $V0$ and $V1$ placed in the vector register $V2$. The vector Y in memory is obtained by storing $V2$ followed by $V3$ back in memory. If Y is the location of the output vector, we do this with the following instructions.

$$\bullet, Y, \quad 1 \quad V2$$

$$\bullet, Y + 3, \quad 1 \quad V3.$$

The computation $y = (A \otimes I_3) P(6, 2) x$ offers a more complicated example. The first step is to load x into two vector registers at stride 2.

$$VO = \begin{bmatrix} x_0 \\ x_2 \\ x_4 \end{bmatrix}, V_1 = \begin{bmatrix} x_1 \\ x_3 \\ x_5 \end{bmatrix}.$$

The next step is

$$V2 = V0 + V1 = \begin{bmatrix} x_0 + x_1 & x_2 + x_3 & x_4 + x_5 \end{bmatrix},$$

$$V3 = V0 - V1 = \begin{bmatrix} x_0 - x_1 & x_2 - x_3 & x_4 - x_5 \end{bmatrix}.$$

Finally the contents of $V2$ are stored at stride 1 beginning at Y and the contents of $V3$ are stored at stride 1 beginning at $Y + 3$. In effect, the stride permutation $P(6, 2)$ is implemented for free since in order to program $A \otimes I_3$ as a vector operation, we must load the input vectors and store the results even in the absense of an input permutation.

The operation $y = P(6,3)(A \otimes I_3)x$ can be performed in the same way.

Implementing a tensor product operation becomes significantly more difficult if segmentation is required, i.e., vectors operations are required on vectors that do not fit inside vector registers. To be concrete, assume that the maximum size of vector registers is 64, and we would like to evaluate $A \otimes I_{128}$. This acts naturally on vectors of size 128. The maximum size of the vector registers is 64, and we would like to replace $A \otimes I_{128}$ by a vector operation on vectors of size 64. Since

$$A \otimes I_{128} = P(256, 128)(I_2 \otimes A \otimes I_{64})P(256, 2),$$

the computation of $A \otimes I_{128}$ is equivalent to $I_2 \otimes A \otimes I_{64}$ upto input and output permutations. Two copies of a vector instruction on vectors of size 64 are required. But $P(256, 2)$ naturally forms vectors of size 128,

$$\begin{bmatrix} x_0 \\ x_2 \\ \vdots \\ x_{254} \end{bmatrix}, \begin{bmatrix} x_1 \\ x_3 \\ \vdots \\ x_{255} \end{bmatrix}.$$

This problem can be solved by the factorization,

$$P(256, 2) = (P(4, 2) \otimes I_{64})(I_2 \otimes P(128, 2)).$$

The first factor $I_2 \otimes P(128, 2)$ decomposes the input vector of size 256 into two consecutive segments of size 128, and performs the stride permutation $P(128, 2)$ on each of the segment.

$$V0 = \begin{bmatrix} x_0 \\ x_2 \\ \vdots \\ x_{126} \end{bmatrix}, V1 = \begin{bmatrix} x_1 \\ x_3 \\ \vdots \\ x_{127} \end{bmatrix}, V2 = \begin{bmatrix} x_{128} \\ x_{130} \\ \vdots \\ x_{254} \end{bmatrix}, V3 = \begin{bmatrix} x_{129} \\ x_{131} \\ \vdots \\ x_{255} \end{bmatrix}.$$

Setting $VL = 64$, these vectors can be loaded by the instructions,

•V0   X,   2

$$\bullet V1 \quad X+1, \quad 2$$

$$\bullet V2 \quad X+128, \quad 2$$

$$\bullet V3 \quad X+129, \quad 2.$$

The second factor $P(4,2) \otimes I_{64}$ can be thought of as a permutation of the segments giving

$$V0 \quad V2 \quad V1 \quad V3.$$

These two steps can be combined by carrying out the load-strides as before, and by changing the initial offsets of the load-strides to perform the permutations of the segments.

The vector operation $A \otimes I_{64}$ is performed on $(V0, V2)$ and on $(V1, V3)$.

$$V4 = V0 + V2$$

$$V5 = V0 - V2$$

$$V6 = V1 + V3$$

$$V7 = V1 - V3.$$

To complete the computation, the vectors $V4$, $V5$, $V6$ and $V7$ must be stored back in the memory in the order given by $P(256, 128)$. This can be done with the store instructions by first permuting the segments to

$$V4 \quad V6 \quad V5 \quad V7$$

and then storing the results with the store instructions

$$\bullet, Y, \quad 2 \quad V4$$

$$\bullet, Y+1, \quad 2 \quad V6$$

$$\bullet, Y+128, \quad 2 \quad V5$$

$$\bullet, Y+129, \quad 2 \quad V7.$$

This corresponds to the factorization

$$P(256, 128) = (I_2 \otimes P(128, 64)(P(4, 2) \otimes I_{64}).$$

2.6. Parallel Implementation of Tensor Products

Tensor product identities provide powerful tools for matching tensor product factor computations to specific machine characteristics such as locality and granularity. Consider the tensor product factor $I_M \otimes A$, where A is taken as in section 5. In the simplest case M separate processors are available for the computation and each processor has access to a shared memory containing the input vector X and the output vector Y. (In this section, we will use capital letters as variable names of the data used in codes.) Number the processors

$$0, 1, \ldots, M - 1. \tag{1}$$

Define the action A by

$$Y(0) = X(0) + X(1),$$

$$Y(1) = X(0) - X(1).$$

Assign to each processor the code

$$A(2, Y, X). \tag{2}$$

The m-th processor acts by this code on the components $X(2m)$, $X(2M + 1)$ of the input vector X and places the results in memory as the components $Y(2m)$, $Y(2m + 1)$ in the output vector Y. In the same fashion, $A \otimes I_M$ is computed by having the m-th processor act on the components $X(m)$, $X(m + M)$. The results are placed in memory as the components $Y(m)$, $Y(m + M)$ in the output vector Y.

If the number R of processors is less than M, then the problem is more complicated. Suppose $M = RS$. Using the identity

$$I_M \otimes A = I_R \otimes (I_S \otimes A), \tag{3}$$

we assign the code $I_S \otimes A$ to each processor to perform the computation as above with R replacing M in the discussion. In the same way, the identity

$$A \otimes I_M = P(2M, 2S)(I_R \otimes (A \otimes I_S))P(2M, R) \tag{4}$$

suggests that each processor be assigned the code for $A \otimes I_R$ with addressing determined by the input and output stride permutations. In particular, the m-th processors, $0 \le m < R$, performs $A \otimes I_S$ on the $2S$ components

$$X(m), X(m + R), \ldots, X(m + (2S - 1)R), \tag{5}$$

and places the result in memory as the $2S$ components

$$Y(m), Y(m + R), \ldots, Y(m + (2S - 1)R). \tag{6}$$

Alternatively, we can use the identity

$$A \otimes I_M = (P(2R, 2) \otimes I_S)(I_R \otimes A \otimes I_S)(P(2R, R) \otimes I_S). \tag{7}$$

Consider the factor $I_M \otimes A \otimes I_N$. As above, we can implement the action by M parallel computations of $A \otimes I_N$. If MN processors are available, we can use the identity

$$I_M \otimes A \otimes I_N = P(2MN, 2M)(I_{MN} \otimes A)(P(2MN, N)) \tag{8}$$

or the identity

$$I_M \otimes A \otimes I_N = (P(2M, M) \otimes I_N)(I_{MN} \otimes A)(P(2M, 2) \otimes I_N) \tag{9}$$

to compute $I_M \otimes A \otimes I_N$ as MN parallel computations by A. In this way, we naturally control the granularity of the parallel computation and fit the computation to granularity and to the number of available processors. The stride permutations give an automatic addressing to the processors.

Theses ideas can be used to compute the tensor product of, say T, factors of A in parallel. By the fundamental factorization,

$$A \otimes \ldots \otimes A = \prod_{t=1}^{T} (I_{2^{T-t}} \otimes A \otimes I_{2^{t-1}}), \tag{10}$$

we decompose the computation into a sequence of computations which at the $(T - t)$-th stage is given by

$$I_{2^{T-t}} \otimes A \otimes I_{2^{t-1}}. \tag{11}$$

To carry out the computation in this way requires a barrier synchronization to guarantee that the input to the next stage is correct. The natural interpretation of each stage leads to a different degree of parallelism at each stage. The factorization must be modified by different addressing, and hence different programming at each stage is required to get a consistent degree of parallelism. We turn to the factorization,

$$A \otimes \ldots, \otimes A = \prod_{t=1}^{T} P(2^T, 2)(I_{2^{T-1}} \otimes A), \tag{12}$$

given in section 2. The addressing is the same at each stage and the natural interpretation has the maximal degree of parallelism at each stage. For example,

$$A \otimes A \otimes A = \prod_{t=1}^{3} P(8, 2)(I_4 \otimes A), \tag{13}$$

and at each stage of the computation, we compute

$$\underline{y} = P(8, 2)(I_4 \otimes A)\underline{x}. \tag{14}$$

We compute this as

$$Y(0) = X(0) + X(1)$$

$$Y(4) = X(0) - X(1)$$

$$Y(1) = X(2) + X(3)$$

$$Y(5) = X(2) - X(3) \tag{15}$$

$$Y(2) = X(4) + X(5)$$

$$Y(6) = X(4) - X(5)$$

$$Y(3) = X(6) + X(7)$$

$$Y(7) = X(6) - X(7).$$

If four processors are available, the the m-th processor $0 \le m < 4$, computes

$$Y(m) = X(2m) + X(2M + 1),$$

$$Y(m + 4) = X(2m) - X(2M + 1). \tag{16}$$

Suppose we have two parallel processors. Then we rewrite (14) as

$$y = P(8, 2)(I_2 \otimes (I_2 \otimes A)) \tag{17}$$

and compute the first four lines of (15) on the 0-th processor and the second four lines of (15) on the first processor. Thus, on the m-th processor, $0 \le m < 2$, we compute

$$\text{for } n = 0, 1. \tag{18}$$

$$Y(2m + n) = X(4m + 2n) + X(4m + 2n + 1),$$

$$Y(2m + n + 4) = X(4m + 2n) - X(4m + 2n + 1)$$

Suppose we wish to compute

$$\bigotimes_{t=1}^{10} A = \prod_{t=1}^{10} P(2^{10}, 2)(I_{2^9} \otimes A), \tag{19}$$

with 8 processors. Writing

$$P(2^{10},2)(I_{2^9} \otimes A) = P(2^{10},2)(I_{2^3} \otimes (I_{2^6} \otimes A)), \qquad (20)$$

at each stage the m-th processor, $0 \le m < 7$, computes

$$\text{for } n = 0, \ldots, 63 \qquad (21)$$

$$Y(2^7 m + n) = X(2^7 m + 2n) + X(2^7 m + 2n + 1)$$

$$Y(2^7 m + n + 2^9) = X(2^7 m + 2n) - X(2^7 m + 2n + 1).$$

In this example, 10 passes are required. After each computational stage, the results are stored back to main (shared) memory. It may be advantageous to do more computations before doing the memory operation. We can do this by the factorization

$$\bigotimes_{t=1}^{10} A = \prod_{t=1}^{5} P(2^{10},2^2)(I_{2^5} \otimes (I_{2^5} \otimes A \otimes A)). \qquad (22)$$

There are only 5 computational stages, reducing transfers to main memory. However, the granularity has been increased since A has been replaced by $A \otimes A$.

In section 3, we discussed the importance of stride permutation factorizations in implementation. For example,

$$P(2^{10},2) = (P(2^4,2) \otimes I_{2^6})(I_{2^3} \otimes P(2^7,2)). \qquad (23)$$

In the case of 8 processors, $I_{2^3} \otimes P(2^7,2)$ is carried out by permuting elements in local memory for each of the processors by $P(2^7,2)$. The results can then be transfered to main memory in segments of length 2^6 permuted by $P(2^4,2)$. In this way the transfer to main memory given by $P(2^{10},2)$ is replaced by decomposing this transfer into a collection of local permutations followed by a global block permutation.

[References]

[1] Johnson, J., Johnson, R., Rodriguez, D. and Tolimieri, R. "A Methodology for Designing, Modifying, and Implementing Fourier Transform Algorithms on Various Architectures." Accepted for publication by *Circuits, Systems and Signal Processing.*

[2] Hoffman, K. and Kunze, R. *Linear Algebra*, Second Ed. Prentice Hall. 1971.

[3] Nering, E. D. *Linear Algebra and Matrix Theory*, John Wiley & Sons. 1970

Problems

1. Show that the tensor product of vectors is bilinear.

2. Show that the tensor product of matrices bilinear.

3. Compute $(A \otimes B)(\underline{a} \otimes \underline{b})$ for

$$A = \begin{bmatrix} 2 & 1 \\ 1 & 1 \end{bmatrix}, B = \begin{bmatrix} 0 & 1 & 0 \\ 2 & 1 & 1 \\ 0 & 1 & 1 \end{bmatrix}$$

$$\underline{a} = \begin{bmatrix} 1 \\ 0 \end{bmatrix}, \underline{b} = \begin{bmatrix} 2 \\ 3 \end{bmatrix}.$$

4. For vectors $\underline{a}$ and $\underline{b}$ running over all vectors of sizes 2 and 3, respectively, show that the tensor products $\underline{a} \otimes \underline{b}$ span $\mathbb{C}^6$.

5. Show the general result: For vectors $\underline{a}$ and $\underline{b}$ running over all vectors of sizes L and M, respectively, the tensor products $\underline{a} \otimes \underline{b}$ span $\mathbb{C}^M$.

6. The *canonical basis* of $\mathbb{C}^L$ is the set of vectors given by the vectors of size L,

$$\underline{e}_0^{(L)} = \begin{bmatrix} 1 \\ 0 \\ 0 \\ \vdots \\ 0 \end{bmatrix}, \underline{e}_1^{(L)} = \begin{bmatrix} 0 \\ 1 \\ 0 \\ \vdots \\ 0 \end{bmatrix}, \underline{e}L - 1^{(L)} = \begin{bmatrix} 0 \\ 0 \\ \vdots \\ 0 \\ 1 \end{bmatrix}.$$

Show that the LM tensor products

$$\underline{e}_r^{(L)} \otimes \underline{e}_s^{(M)} \quad 0 \le r < L,\ 0 \le s < M$$

describe the canonical basis of $\mathbb{C}^{LM}$. (Explicitly derive the $\underline{e}_r^{(L)} \otimes \underline{e}_s^{(M)}$).

7. Describe $P(27,3)$, $P(27,9)$ and show that $P(27,3)P(27,9) = I_{27}$.

8. Compute the matrix product $P(12,2)P(12,3)$.

9. Show that the set $\{P(81,3^s) \mid s/81\}$ is a cyclic group of order 4.

10. Compute $P(8,2)(A \otimes B)P(8,4)$, where

$$A = \begin{bmatrix} 1 & 0 & 0 & 0 \\ 0 & 0 & 1 & 0 \\ 1 & 0 & 0 & 0 \\ 0 & 0 & 0 & 1 \end{bmatrix}, B = \begin{bmatrix} 1 & 1 \\ 1 & -1 \end{bmatrix}.$$

and show that it is equal to $B \otimes A$.

Chapter 3

COOLEY-TUKEY FFT ALGORITHMS

3.1. <u>Introduction</u>

In the following two chapters, we will concentrate on algorithms for computing FFT of size a composite number N. The main idea is to use the additive structure of the indexing set Z/N to define mappings of the input and output data vectors into 2-dimensional arrays. Algorithms are then designed, transforming 2-dimensional arrays which, when combined with these mappings, compute the N-point FFT. The stride permutations of chapter 2 play a major role.

Historically, the first additive FFT algorithm is described in the fundamental work of J. W. Cooley and J. W. Tukey [2] in 1965. Straightforward computation of N-point FFT requires a number of arithmetic operations proportional to N^2. In scientific and technological applications, the transform size N is commonly too large for direct digital computer implementation. The Cooley-Tukey FFT algorithm significantly reduced the computational cost, for many transform sizes N, to an operational count proportional to $N \log N$. This result set the stage for widespread advances in digital hardware, and is one of the main reasons that digital computation has become the overwhelmingly preferred method for computing the FT in most scientific and technological applications.

The years following publication of the Cooley-Tukey FFT saw various implementations of the algorithm on sequential machines[1]. Recently, however, as vector and parallel computer architectures began to play increasingly important roles in scientific computations, the adaptation of the Cooley-Tukey FFT and its variants to these new architectures has become a major research effort. The tensor product provides the key mathematical language in which to describe and analyze, in a unified format, similarities and differences among these algorithms. An account of these variants not using this language can be found in [8]. In 1968, M. Pease [5] utilized the language of the tensor products to formulate a variant of the algorithm which is suitable for implementation on a special purpose parallel computer. In 1983, C. Temperton [9] provided tensor product formulations of the most commonly known variants.

One of the advantages of using tensor product language to describe FFT algorithms is that this mathematical language may be used as an analytic tool for the study of algorithmic structures for machine hardware and software implementations as well as the identification of new algorithms. For instance, an inherent part in the study of computer implementation of FFT algorithms is the analysis of the data communication aspects of the algorithms which manifest themselves during implementation procedures. These data communication aspects can be best studied, in turn, through the analysis of the permutation matrices, the stride permutation, which appear in our tensor product formulation of the FFT algorithms.

We present, in tensor product form, the description of FFT algorithms with the following objective in mind: To provide the user of these algorithms with guidelines which will enable him to effectively study their implementation on either special purpose or general purpose computers. By "effectively studying their implementation," we mean to be able to produce algorithms which best conform to the inherent constraints on any given machine hardware architecture.

In this chapter, we consider the Cooley-Tukey FFT algorithm corresponding to the decomposition of the transform size N into the product of two factors. The convention introduced in chapter 2 relating 2-dimensional arrays to one-dimensional arrays will still be enforced: If X is an $M \times N$ matrix, then we associate to X the MN-tuple $\underline{x}$ formed by reading, in order, down the columns of X. In the sections that follow, algorithms will be designed by using the additive structure of the indexing set to associate a 2-dimensional array to a 1-dimensional array.

3.2. Basic Properties of FFT Matrix

The FFT *matrix of order* N, denoted by $F(N)$ is defined as

$$F(N) \; = \; \left[w_N^{jk} \right], \quad w_N = exp(2\pi i/N), \; i = \sqrt{-1}. \tag{1}$$

The *conjugate* of w_N, denoted by w_N^* is

$$(w_N)^* \; = \; e(-2\pi i/N) \; = \; w_N^{-1}, \tag{2}$$

and

$$(w_N^k)^* \; = \; w_N^{-k \bmod N} \; = \; w_N^{N-k} \; = \; (w_N^k)^*. \tag{3}$$

Direct computation shows that

$$F(N)F(N)^* \; = \; N I_N. \tag{4}$$

The *inverse* FFT matrix is

$$F(N)^{-1} \; = \; \frac{1}{N} F(N)^*, \tag{5}$$

and $F(N)$ is *symmetric*, i.e.,

$$F(N)^t = F(N). \tag{6}$$

3.3. An Example of FFT Algorithm

The 8-point FFT is given by the formula

$$y_k \; = \; \sum_{j=0}^{7} w^{jk} x_j, \quad 0 \le k < 8, \; w = exp(2\pi i/8). \tag{1}$$

Associate to the input vector $\underline{x}$, the 4×2 array

$$X = \begin{bmatrix} x_0 & x_4 \\ x_1 & x_5 \\ x_2 & x_6 \\ x_3 & x_7 \end{bmatrix} \tag{2}$$

and set

$$X_1 = X^t = \begin{bmatrix} x_0 & x_1 & x_2 & x_3 \\ x_4 & x_5 & x_6 & x_7 \end{bmatrix}. \tag{3}$$

The vector $\underline{x}_1$ corresponding to X_1 can be obtained by

$$\underline{x}_1 = P(8,4)\underline{x}, \tag{4}$$

where $P(8,4)$ is the 8-point stride 4 permutation.

Associate to the output vector $\underline{y}$ the 2×4 array

$$Y = \begin{bmatrix} y_0 & y_2 & y_4 & y_6 \\ y_1 & y_3 & y_5 & y_7 \end{bmatrix}. \tag{5}$$

We will rewrite (1) in terms of the arrays X_1 and Y. First

$$X_1(k_1, k_2) = x(k_2 + 4k_1), \quad 0 \le k_1 < 2, \ 0 \le k_2 < 4, \tag{6}$$

$$Y(l_1, l_2) = y(l_1 + 2l_2), \quad 0 \le l_1 < 2, \ 0 \le l_2 < 4. \tag{7}$$

Placing these formulas into (1), we have

$$Y(l_1, l_2) = \sum_{k_2=0}^{3} \left[\sum_{k_1=0}^{1} X_1(k_1, k_2) w^{(k_2 + 4k_1)(l_1 + 2l_2)} \right]. \tag{8}$$

Set $v = w^2 = i$ and $u = w^4 = 1$. Then

$$w^{(k_2 + 4k_1)(l_1 + 2l_2)} = u^{k_1 l_1} v^{k_2 l_2} w^{k_2 l_1}. \tag{9}$$

We can rewrite (8) as

$$Y(l_1, l_2) = \sum_{k_2=0}^{3} \left[\sum_{k_1=0}^{1} X_1(k_1, k_2) u^{k_1 l_1} \right] w^{k_2 l_1} v^{k_2 l_2}. \tag{10}$$

We can decompose (10) into a sequence of operations as follows. First we compute the inner sum

$$Y_1(l_1, l_2) = \sum_{k_1=0}^{1} X_1(k_1, k_2) u^{k_1 l_1}, \quad 0 \le l_1 < 2, \ 0 \le k_2 < 4. \quad (11)$$

We see from (11) that the array Y_1 is computed by taking the 2-point FFT of each column of the array X_1. In tensor notation

$$\underline{y}_1 = (I_4 \otimes F(2))\underline{x}_1, \quad (12)$$

where $\underline{y}_1$ is the vector corresponding to Y_1.

The next stage of the computation

$$Y_2(l_1, k_2) = Y_1(l_1, k_2) w^{l_2 k_1}, \quad (13)$$

introduces the *twiddle factor*. In matrix notation,

$$\underline{y}_2 = T\underline{y}_1, \quad (14)$$

where $\underline{y}_2$ is the vector corresponding to Y_2 and T is the diagonal matrix

$$T = diag(1, 1, 1, w, 1, w^2, 1, w^3). \quad (15)$$

We complete the computation of Y from (11) by

$$Y(l_1, l_2) = \sum_{k_2=0}^{3} Y_2(l_1, k_2) v^{k_2 l_2}, \quad 0 \le l_1 < 2, \ 0 \le l_2 < 4, \quad (16)$$

which is given by the 4-point FFT of each row of the array Y_2. In tensor notation (16) becomes

$$\underline{y} = (F(4) \otimes I_2)\underline{y}_2. \quad (17)$$

Combining these formulas, we have

$$\underline{y} = (F(4) \otimes I_2)T(I_4 \otimes F(2))P(8, 4)\underline{x}. \quad (18)$$

This leads to the factorization

$$F(8) = (F(4) \otimes I_2)T(I_4 \otimes F(2))P(8,4). \tag{19}$$

3.4. Cooley-Tukey FFT for $N = 2M$

The N-point FFT is given by the formula

$$y_k = \sum_{j=0}^{N-1} w^{jk}x_j, \quad 0 \le k < N, \ w = exp(2\pi i/N). \tag{1}$$

Associate to the N-point input data $\underline{x}$ the $M \times 2$ array

$$X = \begin{bmatrix} x_0 & x_M \\ x_1 & x_{M+1} \\ \vdots & \\ x_{M-1} & x_{N-1} \end{bmatrix}, \tag{2}$$

and set

$$X_1 = X^t = \begin{bmatrix} x_0 & x_1 & \cdots & x_{M-1} \\ x_M & x_{M+1} & \cdots & x_{N-1} \end{bmatrix}. \tag{3}$$

The corresponding N-tuple $\underline{x}_1$ is given by

$$\underline{x}_1 = P(N,M)\underline{x}. \tag{4}$$

Associate to the N-point output data $\underline{y}$ the $2 \times M$ array

$$Y = \begin{bmatrix} y_0 & y_2 & \cdots & y_{N-2} \\ y_1 & y_3 & \cdots & y_{N-1} \end{bmatrix}. \tag{5}$$

We can rewrite (1) in terms of the two-dimensional arrays X_1 and Y. First

$$X_1(k_1, k_2) = x(k_2 + Mk_1), \quad 0 \le k_1 < 2, 0 \le k_2 < M, \tag{6}$$

and

$$Y(l_1, l_2) = y(l_1 + 2l_2), \quad 0 \le l_1 < 2, 0 \le l_2 < M. \tag{7}$$

Using (6) and (7), we have

$$Y(l_1, l_2) = \sum_{k_2=0}^{M-1} \left(\sum_{k_1=0}^{1} X_1(k_1, k_2) u^{k_1 l_1} \right) w^{k_2 l_1} v^{k_2 l_2}, \qquad (8)$$

where $v = w^2$, $u = w^M = -1$ and $w^N = 1$. The inner sum,

$$Y_1(l_1, k_2) = \sum_{k_1=0}^{1} X_1(k_1, k_2) u^{k_1 l_1}, \qquad (9)$$

computes, for each $0 \le k_2 < M$, the 2-point FFT of the corresponding column of X_1. Let $\underline{x}_1$ be the vector associated to the 2-dimensional array X_1. To compute (7), we partition $\underline{x}_1$ into m vectors each of length 2 given by the columns of X_1, and compute the 2-point FFT of these vectors. The ouput is placed in the columns of Y_1. In tensor product notation

$$\underline{y}_1 = (I_m \otimes F(2)) P \underline{x}, \quad P = P(N, M). \qquad (10)$$

There are two remaining steps in the computation. First, we compute

$$Y_2(l_1, k_2) = Y_1(l_1, k_2) w^{l_2 l_1}, \qquad (11)$$

which can be described by the diagonal matrix multiplication

$$\underline{y}_2 = T \underline{y}_1, \qquad (12)$$

where

$$T = diag(1\, 1\, 1w \ldots 1 w^{M-1}). \qquad (13)$$

The final computation

$$Y(l_1, l_2) = \sum_{k_2=0}^{M-1} Y_2(l_1, k_2) v^{k_2 l_2}, \qquad (14)$$

computes the M-point FFT of the rows of Y_2 which can be written as

$$\underline{y} = (F(M) \otimes I_2) \underline{y}_2. \qquad (15)$$

This discussion leads to the following theorem

__Theorem__ 1 Let $N = 2M$. Then

$$F(N) = (F(M) \otimes I_2)T(I_M \otimes F(2))P(N, M), \qquad (16)$$

where T is the diagonal matrix (13).

The permutation $P(N, M)$ naturally forms vectors of size 2 on which the action of $I_M \otimes F(2)$ can be computed in parallel. The twiddle factor T can be thought of as a block diagonal matrix consisting of M diagonal blocks of size 2. Each of the M blocks acts, in parallel, on the 2-dimensional vector resulting from the action of $I_M \otimes F(2)$. The computation is completed by the vectorized FT, $F(M) \otimes I_2$; the vector FT $F(M)$ acting on the set of M 2-dimensional vectors. For example, if $N = 2 \cdot 3$, then the computation of $F(6)$ can be represented by the following diagram.

3.5. Twiddle Factors

In this section, we consider the twiddle factors or the diagonal matrices appearing in Cooley-Tukey FFT algorithms.

For $N > 1$, set

$$D(N) = \begin{bmatrix} 1 & & & \\ & w & & \\ & & \ddots & \\ & & & w^{N-1} \end{bmatrix}, \quad w = exp(2\pi i/N), \qquad (1)$$

and for $1 \leq r \leq N$

$$D_r(N) = \begin{bmatrix} 1 & & & \\ & w & & \\ & & \ddots & \\ & & & w^{r-1} \end{bmatrix}. \qquad (2)$$

__Example__ 1. Take $N = 2$. Then

$$D(2) = \begin{bmatrix} 1 & 0 \\ 0 & -1 \end{bmatrix}.$$

Example 2. Take $N = 4$. Then

$$D(4) = \begin{bmatrix} 1 & & & \\ & i & & \\ & & -1 & \\ & & & -i \end{bmatrix},$$

and

$$D_2(4) = \begin{bmatrix} 1 & 0 \\ 0 & i \end{bmatrix}.$$

There are important formulas relating these diagonal matrices. First

$$D(4) = D(2) \otimes D_2(4). \tag{3}$$

Assume $N = 2M$. With $w = exp(2\pi i/N)$ and $w^2 = exp(2\pi i/M)$,

$$D(N) = \begin{bmatrix} D_2(N) & & & \\ & w^2 D_2(N) & & \\ & & \ddots & \\ & & & w^{2(M-1)} D_2(N) \end{bmatrix}. \tag{4}$$

By the definition of the tensor product

$$D(N) = D(M) \otimes D_2(N). \tag{5}$$

The general result will be stated as a theorem.

Theorem 1. If $N = RS$, then

$$D(N) = D(S) \otimes D_R(N). \tag{6}$$

Proof First, notice that

$$w^R = e(2\pi i/S). \tag{7}$$

Thus we can write

$$D(N) = \begin{bmatrix} D_R(N) & & & \\ & w^R D_R(N) & & \\ & & \ddots & \\ & & & w^{R(S-1)} D_R(N) \end{bmatrix} \tag{8}$$

which by definition proves (6).

Let $N = RS$. Define the matrix direct sum

$$T_R(N) = \sum_{s=0}^{S-1} \oplus D_R^s(N)$$

$$= \begin{bmatrix} I_R & & & & \\ & D_R(N) & & & \\ & & \cdot & & \\ & & & \cdot & \\ & & & & D_R^s(N) \end{bmatrix}. \qquad (9)$$

The matrix $T_R(N)$ can be viewed as a block diagonal matrix consisting of S diagonal blocks of size R. In this way, it naturally acts, in parallel, on S vectors of size R. The diagonal matrix T of theorem 4.1 is $T_2(N)$.

Stride permutations act on these diagonal matrices as follows.

Theorem 2. If $N = RS$, then

$$P(N, R)T_R(N)P(N, S) = \sum_{r=0}^{R-1} \otimes D_S^r(N) = T_S(N). \qquad (10)$$

Proof The matrix on the left hand side of (10) is the diagonal matrix formed from the product of $P(N, R)$ with the vector formed from the diagonal components of $T_R(N)$. Listing the diagonal elements of $T_R(N)$ in a row,

$$1\,1\,\ldots\,1; 1\,w\,\ldots w^{R-1}; \ldots; 1\,w^{S-1}\,\ldots\,w^{(S-1)(R-1)} \qquad (11)$$

and striding through the row with stride R, we have

$$1\,1\,\ldots\,1; 1\,w\,\ldots w^{S-1}; \ldots; 1\,w^{R-1}\,\ldots\,w^{(S-1)(R-1)} \qquad (12)$$

proving the theorem.

As expected, the natural block-like structure of $T_R(N)$ is transformed into R diagonal blocks of size S by conjugation by $P(N, R)$.

The judicious application of this result provides a means of keeping consistency throughout the computation.

3.6. <u>FT Factors</u>

In general, we will need to compute the actions of

$$I_R \otimes F(S), \tag{1}$$

$$F(S) \otimes I_R. \tag{2}$$

Factors of these types were studied in chapter 2. Although the arithmetic cost of both actions is the same, the efficiency of implementation can vary on different types of machine architecture. Diagrams can be helpful. Represent multiplication α by

$$x_0 . \qquad\qquad .\alpha x_0$$

and the computation, $y = \alpha_0 x_0 + \alpha_1 x_1$, by

$$x_0 .$$
$$\qquad\qquad \alpha_0 x_0 + \alpha_1 x_1 . \tag{4}$$
$$x_1 .$$

If $\alpha = 1$, we omit it. The action of $F(2)$ can be represented by the butterfly diagram

$$x_0 . \qquad\qquad .x_0 + x_1$$
$$x_1 . \qquad\qquad .x_0 - x_1 \tag{5}$$

<u>Example</u> 1. The action of $I_2 \otimes F(2)$ is represented by

$$x_0 . \qquad\qquad .x_0 + x_1$$
$$x_1 . \qquad\qquad .x_0 - x_1$$
$$\tag{6}$$
$$x_2 . \qquad\qquad .x_2 + x_3$$
$$x_3 . \qquad\qquad .x_2 - x_3$$

which we see consists of 2 parallel 2-point FFT's.

In general, the action of $I_R \otimes F(R)$ can be computed by r parallel s-point FFT's.

Example 2. The action of $F(2) \otimes I_2$ is represented by

$$
\begin{array}{ll}
x_0. & .x_0 + x_2 \\
x_1. & .x_1 + x_3 \\
x_2. & .x_0 - x_2 \\
x_3. & .x_1 - x_3.
\end{array}
\tag{7}
$$

In the previous chapter, we saw that this is a vector operation. Associate to $\underline{x}$ the equivalent 2-dimensional array

$$
X = \begin{bmatrix} x_0 & x_2 \\ x_1 & x_3 \end{bmatrix}.
\tag{8}
$$

Form the 2 vectors of length 2 from the columns of X

$$
\underline{x}(0) = \begin{bmatrix} x_0 \\ x_1 \end{bmatrix} \quad \underline{x}(1) = \begin{bmatrix} x_2 \\ x_3 \end{bmatrix}.
\tag{9}
$$

We say that these vectors are formed with stride 1. The action of $F(2) \otimes I_2$ is given by the 2-point vector FFT,

$$
\begin{bmatrix} \underline{y}(0) \\ \underline{y}(1) \end{bmatrix} = F(2) \begin{bmatrix} \underline{x}(0) \\ \underline{x}(1) \end{bmatrix}.
\tag{10}
$$

We read out the vector $\underline{y} = (F(2) \otimes I_2)\underline{x}$ from (10) in natural order.

As in chapter 2, the commutation theorem can be used to interchange parallel and vector operations by the formula

$$
P(N,S)(I_R \otimes F(S))P(N,R) = F(S) \otimes I_R, \quad N = RS.
\tag{11}
$$

As an example, observe that the following diagram also computes the action $F(2) \otimes I_2$ (which is a vector operation) as a parallel operation, using the commutation theorem:

Example 3. $F(2) \otimes I_2 = P(4,2)(I_2 \otimes F(2))P(4,2)$:

$$
\begin{array}{cccc}
x_0. & .x_0 & .x_0 + x_2 & .x_0 + x_2 \\
x_1. & .x_2 & .x_0 - x_2 & .x_1 + x_3 \\
x_2. & .x_1 & .x_1 + x_3 & .x_0 - x_2 \\
x_3. & .x_3 & .x_1 - x_3 & .x_1 - x_3
\end{array}
\tag{12}
$$

Example 4. $y = (F(2) \otimes I_3)x$. Then

$$
\begin{array}{ll}
y_0 = x_0 + x_3 \;, & y_3 = x_0 - x_3, \\[4pt]
y_1 = x_1 + x_4 \;, & y_4 = x_1 - x_4, \\[4pt]
y_2 = x_2 + x_5 \;, & y_5 = x_2 - x_5,
\end{array}
\tag{13}
$$

which can be represented by

$$
\begin{array}{cc}
x_0. & .x_0 + x_3 \\
x_1. & .x_1 + x_4 \\
x_2. & .x_2 + x_5 \\
x_3. & .x_0 - x_3 \\
x_4. & .x_1 - x_4 \\
x_5. & .x_2 - x_5
\end{array}
\tag{14}
$$

This computation can be carried out in three stages as indicated by the diagram

$$
\begin{array}{cccc}
x_0. & .x_0 & .x_0 + x_3 & .x_0 + x_3 \\
x_1. & .x_3 & .x_0 - x_3 & .x_1 + x_4 \\
x_2. & .x_1 & .x_1 + x_4 & .x_2 + x_5 \\
x_3. & .x_4 & .x_1 - x_4 & .x_0 - x_3 \\
x_4. & .x_2 & .x_2 + x_5 & .x_1 - x_4 \\
x_5. & .x_5 & .x_2 - x_5 & .x_2 - x_5
\end{array}
\tag{15}
$$

which corresponds to the factorization

$$
F(2) \otimes I_3 = P(6,2)(I_3 \otimes F(2))P(6,3).
\tag{16}
$$

3.7. Variants of Cooley-Tukey FFT Algorithm

In the notation of the preceeding section,

$$F(N) = (F(M) \otimes I_2)T_2(N)(I_M \otimes F(2))P(N, M), \quad N = 2M. \quad (1)$$

We will now derive other factorizations of $F(N)$. These variants are distinguished by the flow of the data through the computation. They make available to the algorithm designer several possibilities for computing N-point FFT, all having the same arithmetic, but differing in storage and gathering of data.

By the commutation theorem,

$$P(N, M)(I_2 \otimes F(M))P(N, 2) = F(M) \otimes I_2. \quad (2)$$

Using this formula in (1), we have

$$F(N)$$

$$= P(N, M)(I_2 \otimes F(M))P(N, 2)T_2(N)(I_m \otimes F(2))P(N, M). \quad (3)$$

Both of the FFT factors are parallel factors; The first is M copies of $F(2)$ and the second is 2 copies of $F(M)$. The first part of the computation, $T_2(N)(I_M \otimes F(2))P(N, M)$, is naturally thought of as parallel action on vectors of size 2, while the second part, $(I_2 \otimes F(M))P(N, 2)$, is a parallel action on vectors of size M.

Applying the commutation theorem in the form

$$P(N, 2)(I_m \otimes F(2))P(N, M) = F(2) \otimes I_M \quad (4)$$

leads to the factorization

$$F(N) = (F(M) \otimes I_2)T_2(N)P(N, M)(F(2) \otimes I_M). \quad (5)$$

The FFT factors are now vector factors; the first acting on vectors of size M and the second acting on vectors of size 2.

A second technique for manipulating factorizations comes from the transpose. Taking the transpose on both sides of (1) and using the formulas

$$F(N)^t = F(N), \quad (6)$$

$$P^{-1} = P^t, \tag{7}$$

$$(A \otimes B)^t = A^t \otimes B^t, \tag{8}$$

results in the factorization

$$F(N) = P(N,2)(I_M \otimes F(2))T_2(N)(F(M) \otimes I_2). \tag{9}$$

A permutation of output data is now required.

Applying transpose and the commutation theorem, other factorizations can be derived. There are several features which distinguish between these factorizations; input permutation, output permutation and internal permutation, the type of lower order FT factors and their placement in the computation. We single out the 4 factorizations derived in this section for future reference: Using theorem 4.1, we have for $N = 2M$,

(a) $\qquad F(N) = (F(M) \otimes I_2)T(I_M \otimes F(2))P,$

(b) $\qquad F(N) = P^{-1}(I_M \otimes F(2))T(F(M) \otimes I_2),$

(c) $\qquad F(N) = P^{-1}(I_2 \otimes F(M))PT(I_M \otimes F(2))P,$

(d) $\qquad F(N) = (F(M) \otimes I_2)TP(F(2) \otimes I_M),$

where $P = P(n,m)$ and

$$T = T_2(N) = \sum_{m=0}^{M} \oplus D_2^m(N), \tag{10}$$

$$D_2(N) = \begin{bmatrix} 1 & 0 \\ 0 & w \end{bmatrix} \quad w = exp(2\pi i/N). \tag{11}$$

3.8. Cooley-Tukey for $N = RS$

Let $N = RS$ and consider the N-point FFT

$$y_k = \sum_{n=0}^{N-1} w^{nk} x_n, \quad 0 \le k < N, \ w = exp(2\pi i/N). \quad (1)$$

We will derive a Cooley-Tukey algorithm computing N-point FFT. The commutation theorem and the transpose is applied to derive other forms.

Associate to the N-point input vector $\underline{x}$ the $S \times R$ array

$$X = \begin{bmatrix} x_0 & x_S & \cdots & x_{R-1)S} \\ x_1 & x_{S+1} & \cdots & x_{(R-1)S+1} \\ \vdots & & & \\ x_{S-1} & x_{2S-1} & \cdots & x_{N-1} \end{bmatrix}, \quad (2)$$

and set

$$X_1 = X^t = \begin{bmatrix} x_0 & x_1 & \cdots & x_{S-1} \\ x_S & x_{S+1} & \cdots & x_{2S-1} \\ \vdots & & & \\ x_{(R-1)S} & x_{(R-1)S+1} & \cdots & x_{N-1} \end{bmatrix}. \quad (3)$$

The corresponding N-tuple $\underline{x}_1$ is given by applying the N-point stride-S permutation $P(N, S)$ to $\underline{x}$.

Associate to the output vector $\underline{y}$ the $R \times S$ array

$$Y = \begin{bmatrix} y_0 & y_R & \cdots & y_{(S-1)R} \\ y_1 & y_{R+1} & \cdots & y_{(S-1)R+1} \\ \vdots & & & \\ y_{R-1} & y_{2R-1} & \cdots & y_{N-1} \end{bmatrix}. \quad (4)$$

We can write

$$X_1(k_1, k_2) = x(k_2 + k_1 S), \quad 0 \le k_1 < R, \ 0 \le k_2 < S, \quad (5)$$

$$Y(l_1, l_2) = y(l_1 + l_2 R), \quad 0 \le l_1 < R, \ 0 \le l_2 < S. \quad (6)$$

Formula (1) can be rewritten as

$$Y(l_1, l_2) = \sum_{k_2=0}^{S-1} \sum_{k_1=0}^{R-1} w^{(k_2+k_1 S)(l_1+l_2 R)} X_1(k_1, k_2). \quad (7)$$

Now

$$(k_2 + k_1 S)(l_1 + l_2 R) \equiv k_2 l_1 + k_1 l_1 S + k_2 l_2 R \bmod N. \tag{8}$$

Set $u = w^S$ and $v = w^R$. Since $w^N = 1$, we can rewrite (7) as

$$Y(l_1, l_2) = \sum_{k_2=0}^{S-1} \left(\sum_{k_1=0}^{R-1} X_1(k_1, k_2) u^{k_1 l_1} \right) w^{k_2 l_1} v^{k_2 l_2}. \tag{9}$$

The argument proceeds as in section 2. First observe that the inner sum

$$Y_1(l_1, k_2) = \sum_{k_1=0}^{R-1} X_1(j_1, j_2) u^{k_1 l_1} \tag{10}$$

computes, for each $0 \le k_2 < S$, the R-point FFT of the k_2-th column of X_1 and places the result in the k_2-th column of Y_1. Let $\underline{y}_1$ be the vector formed by reading, in order, down the columns of Y_1. Then

$$\underline{y}_1 = (I_S \otimes F(R)) \underline{x}_1 \tag{11}$$

and

$$\underline{y}_1 = (I_S \otimes F(R)) P(N, S) \underline{x}. \tag{12}$$

The next stage of the computation

$$Y_2(k_2, j_2) = Y_1(k_1, j_2) w^{j_2 k_1} \tag{13}$$

can be given by the diagonal matrix multiplication

$$\underline{y}_2 = T_R(N) \underline{y}_1. \tag{14}$$

We complete the computation by

$$Y(l_1, l_2) = \sum_{k_2=0}^{S-1} Y_2(l_1, k_2) v^{k_2 l_2}. \tag{15}$$

The computation is taken on rows.

Theorem 1. If $N = RS$, then

$$\underline{y} = F(N)\underline{x} = (F(S) \otimes I_R)T_R(N)(I_S \otimes F(R))P(N, S)\underline{x}. \qquad (16)$$

The corresponding factorization is

$$F(N) = (F(S) \otimes I_R)T_R(N)(I_S \otimes F(R))P. \qquad (17)$$

The first part of the computation,

$$T_R(N)(I_S \otimes F(R))(P(N, S)), \qquad (18)$$

can be viewed as parallel actions on vectors of size R, while the second part, $F(S) \otimes I_R$, is the vector FT on the resulting S-vectors of size R.

The transpose and the commutation theorem can be used to derive other factorization as in section 5. We single out the following list for future reference:

Case $N = 2M$:

$$D_2(N) = \begin{bmatrix} 1 & 0 \\ 0 & w \end{bmatrix}, w = exp(2\pi i/N).$$

$$T_2(N) = \begin{bmatrix} I_2 & & & & \\ & D_2(N) & & & \\ & & \cdot & & \\ & & & \cdot & \\ & & & & \cdot \\ & & & & & D_2(N)^{M-1} \end{bmatrix}$$

(a) $F(N) = (F(M) \otimes I_2)T_2(N)(I_M \otimes F(2))P(N, M).$

(b)
$$F(N) = P(N, 2)(I_2 \otimes F(M))P(N, M)T_2(N)(I_M \otimes F(2))P(N, M).$$

(c) $F(N) = (F(M) \otimes I_2))T_2(N)P(N, M)(F(2) \otimes I_M).$

Case $N = 2M$:

$$D_M(N) = \begin{bmatrix} 1 & & & & \\ & w & & & \\ & & \cdot & & \\ & & & \cdot & \\ & & & & \cdot \\ & & & & & w^{M-1} \end{bmatrix}, w = exp(2\pi i/N).$$

$$T_M(N) = \begin{bmatrix} I_M & 0 \\ 0 & D_M(N) \end{bmatrix}.$$

(a) $F(N) = (F(2) \otimes I_M)T_M(N)(I_2 \otimes F(M))P(N,2).$

(b)
$$F(N) = P(N,2)(I_M \otimes F(2))P(N,M)T_M(N)(I_2 \otimes F(M))P(N,2).$$

(c) $F(N) = (F(2) \otimes I_M))T_M(N)P(N,2)(F(M) \otimes I_2).$

Case $N = RS$:

$$D_R(N) = \begin{bmatrix} 1 & & & & \\ & w & & & \\ & & \cdot & & \\ & & & \cdot & \\ & & & & \cdot \\ & & & & & w^{R-1} \end{bmatrix}, w = exp(2\pi i/N).$$

$$T_R(N) = \begin{bmatrix} I_R & & & & \\ & D_R(N) & & & \\ & & \cdot & & \\ & & & \cdot & \\ & & & & \cdot \\ & & & & & D_R(N)^{S-1} \end{bmatrix}.$$

(a)
$$F(N) = P(N,S)(I_R \otimes F(S))P(N < R)T_R(N)(I_S \otimes F(R))P(N,S).$$

(b)
$$F(N) = P(N,S)(I_R \otimes F(S))P(N,R)T_R(N)(I_S \otimes F(R))P(N,S).$$

(c) $F(N) = (F(S) \otimes I_R)T_R(N)P(N,S)(F(R) \otimes I_S).$

3.9. Arithmetic Cost

The number of arithmetic operations required to carry out a computation is an important part of the cost of the computation and has traditionally occupied the most attention. On modern architecture machines, a large part of the computation time can be spend on data communication; but, there is as yet little general theory measuring this aspect of the overall computation. We gave some general guidelines in the previous sections but much more, specially on specific architecture, remains unanswered. Arithmetic cost is much easier to estimate.

In the class of algorithms listed on section 5., each algorithm has the same arithmetic cost, if we neglect the underlying arithmetic involved in addressing. Consider factorization (5.1). We require an input permutation at 0 (zero) arithmetic cost. Then SR-point FFT must be computed, followed by a diagonal matrix multiplication. In the last stage, since $F(S) \otimes I_R$ and $I_R \otimes F(S)$ are the same, up to data permutation, the equivalent of RS-point FFT's must be computed. If we have some algorithm computing M-point FFT whith $m(M)$ multiplications and $a(M)$ additions, then the algorithms of section 5 compute N-point FFT using

$$Ra(S) + Sa(R) \qquad (1)$$

additions and

$$Rm(S) + N + Sm(R) \qquad (2)$$

multiplications. The N in (2) comes from the diagonal matrix multiplication. Since many of the diagonal entries are 1 in practice, we can reduce this cost.

If we take

$$a(M) = M(M-1), \tag{3}$$

$$m(M) = M^2, \tag{4}$$

then (1) becomes

$$N(R+S-2) \tag{5}$$

which should be compared to $m(N) = N^2$.

[References]

[1] Cochran, W. T. et al., "What is the Fast Fourier Transform?," IEEE Trans. Audio Electroacoust., vol. 15, 1967, pp.45-55.

[2] Cooley, J. W., Tuckey, J. W. "An Algorithm for the machine Calculation of complex Fourier Series," Math. Comp., vol. 19, 1965, pp.297-301.

[3] Gentleman, W. M.,Sande, G. "Fast Fourier Transform for Fun and Profit," Proc.AFIPS, Joint Computer Conference, vol.29,1966, pp.563-578.

[4] Korn, D.J., Lambiotte, J.J. "Computing the Fast Fourier Transform on a vector computer," Math. Comp., vol.33, 1979, pp.977-992.

[5] Pease, M. C. "An Adaptation of the Fast Fourier Transform for Parallel Processing," J. ACM, vol8, 1971, pp.843-846.

[6] Burrus, C.S. "Bit Reverse Unscrambling for a Radix 2^m FFT", IEEE Trans. on ASSP, vol. 36, July, 1988.

[7] Singleton, R. C. "An Algorithm for Computing the Mixed-Radix Fast Fourier Transform," IEEE Trans.Audio Electroacoust., vol.17, 1969, pp.93-103.

[8] Swartztrauber, P. N. "FFT algorithms for vector computers," Parallel Computing, vol.1, North Holland, 1984, pp.45-63.

[9] Temperton, C. "Self-Sorting Mixed-Radix Fast Fourier Transforms," J. of Compt.Phys., 52(1), 1983 pp.198-204.

[10] Burrus, C.S. and Park, T.W. *DFT/FFT and Convolution Algorithms* New York: John Wiley and Sons, 1985.

[11] Oppenheim, A.V. and Schafer, R.W. *Digital Signal Processing*, Englewood Cliffs, NJ: Prentice-Hall, 1975.

[12] Nussbaumer, H.J. *Fast Fourier Transform and Convolution Algorithms*, Berlin, Heidelberg and New York, Springer-Verlag, 1981.

Problems

1. Show $F(N)F(N)^* = N I_N$.

2. Compute $T_4(16)$, $T_4(64)$ and $T_4(128)$.

3. Compute $T_3(9)$, $T_3(27)$ and $T_3(81)$.

4. Show directly that

$$P(12,4)T_4(12)P(12,3) = T_3(12).$$

5. Diagram the computation of $F(3) \otimes I_4$ using the identity

$$F(3) \otimes I_4 = P(12,3)(I_4 \otimes F(3))P(12,4).$$

6. Derive directly the factorization

$$F(27) = (F(9) \otimes I_3)T_3(27)(I_9 \otimes F(3))P(27,9).$$

7. From theorem (8.1), use the traspose to derive the factorization

$$F(N) = P(N,S)(I_R \otimes F(S))P(N,R)T_R(N)(I_S \otimes F(R))P(N,S).$$

8. Prove $(A \otimes B)^t = A^t \otimes B^t$.

Chapter 4

VARIANTS OF FFT ALGORITHM AND

THEIR IMPLEMENTATIONS

4.1. Introduction

In chapter 3, additive FFT algorithms were derived corresponding to the factorization of the transform size N into the product of
two factors. Analogous algorithms will now be designed corresponding to transform sizes given as the product of three or more factors.
In general, as the number of factors increases, the number of possible
algorithms increases.

In this chapter, we derive the Cooley-Tukey [3] and Gentleman-
Sande [4] FFT algorithms. They are related by matrix transpose,
and distinguished by whether bit-reversal is applied at input or output. In any case, FT factors of mixed-type,

$$I_R \otimes F(T) \otimes I_S \tag{1}$$

appear in the factorization as discussed in chapter 2. The factor
(1) can be viewed as R concurrent FFT's on vectors of length S.
Applying the commutation theorem, this factor can be replaced by
the 'vector' factor

$$F(T) \otimes I_{RS}, \tag{2}$$

which can be viewed as the vector T-point FT on vectors of length
RS. In theory, a vectorization of the Cooley-Tukey FFT algorithm
is produced by systematically replacing all mixed-type FT factors by
their corresponding vector factors. However, implementing a vector
factor on a specific vector computer cannot, in general, be accom-
plished without breaking up the computation into pieces which can
be fit into the vector registers. This partitioning of the computation

introduces concurrency back into the factor, and is one of the main difficulties in matching algorithm to architecture. This problem was discussed in chapter 2.

Parallel algorithms and vector algorithms are easily related by the commutation theorem. The commutation theorem introduces explicit permutation matrices into the factorization. Not surprisingly, these permutation matrices are built from the stride permutations. Variants of the Cooley-Tukey FFT algorithms, to a large extent, depend on which permutation matrices are used to bring about vectorization or parallelization. For example, for $N = RST$, we have the two formulas

$$F(T) \otimes I_{RS} = P(N, ST)(I_R \otimes F(T) \otimes I_S)P(N, ST)^{-1}, \qquad (3)$$

and

$$F(T) \otimes I_{RS}$$

$$= (P(RT, T) \otimes I_S)(I_R \otimes F(T) \otimes I_S)(P(RT, T)^{-1} \otimes I_S). \qquad (4)$$

The variants derived by Pease [6], Korn-Lambiotte [5], and Agarwal-Cooley [1] depend on factorization (3) while the auto-sort variant derived by Stockham and found in [9] depends on factorization (4). The Korn-Lambiotte FFT algorithm is the vectorized analogue of the parallel FFT algorithm of Pease. Two features distinguish these algorithms. First, the main computational stages are the same in all these algorithms and are given by the vector FT factors (2). In the Korn-Lambiotte FFT algorithm and the Agarwal-Cooley FFT algorithm, bit-reversal is required at input or output which can be a time consuming step on many vector computers. However, the internal permutations introduced by the commutation theorem, as seen by (3), have uniform structures throughout the different stages of the computation, and can be implemented by the stride-load memory feature of many vector computers on vectors of maximal length. The auto-sort variant does not require bit-reversal at input or output. It accomplishes this savings by distributing bit-reversal throughout the computation. From (4), we see that the internal permutations are tensor products and can be viewed as vector stride permutations; but, the vector lengths are not maximal and change throughout the computation.

In section 2, we derive the radix-2 Cooley-Tukey FFT algorithm and the radix-2 Gentleman-Sande FFT algorithm. Bit-reversal is defined. In section 3, the Pease FFT algorithm is derived and its vector form due to Korn and Lambiotte is discussed. The auto-sort FFT algorithm is derived in section 4. In the final three sections, the mixed radix generalizations of these algorithms are given.

4.2. Radix-2 Cooley-Tukey FFT Algorithm

We derive a Cooley-Tukey FFT algorithm for transform size $N = 2^k$. The algorithm decomposes the computation of N-point FFT into a sequence of k operations each requiring 2-point FFT's followed by an output permutation. Our derivation is based on the factorization

$$F(N) = P(N)(I_2 \otimes F(M))T(N)(F(2) \otimes I_M), \quad M = N/2, \quad (1)$$

where $P(N) = P(N, M)$ and $T(N) = T_M(N)$. We begin with two examples.

<u>Example</u> 1. Take $N = 2 \times 2$. By (1),

$$F(4) = P(4)(I_2 \otimes F(2))T(4)(F(2) \otimes I_2). \qquad (2)$$

<u>Example</u> 2. Take $N = 2 \times 4$. By (1),

$$F(8) = P(8)(I_2 \otimes F(4))T(8)(F(2) \otimes I_4). \qquad (3)$$

The operation $I_2 \otimes F(4)$ can be factored using (2) and the tensor product identity,

$$I \otimes (BC) = (I \otimes B)(I \otimes C), \qquad (4)$$

with the result

$$I_2 \otimes F(4) = (I_2 \otimes P(4))(I_4 \otimes F(2))(I_2 \otimes T(4))(I_2 \otimes F_2 \otimes I_2). \quad (5)$$

Placing (5) into (3), we have the factorization of $F(8)$,

$$P(8)(I_2 \otimes P(4))(I_4 \otimes F(2))(I_2 \otimes T(4))(I_2 \otimes F(2) \otimes I_2)T(8)(F(2) \otimes I_4). \quad (6)$$

We organize the computation into stages by setting

$$X_1 = T(8)(F(2) \otimes I_4), \quad (7)$$

$$X_2 = (I_2 \otimes T(4))(I_2 \otimes F(2) \otimes I_2), \quad (8)$$

$$X_3 = I_4 \otimes F(2), \quad (9)$$

$$Q = P(8)(I_2 \otimes P(4)). \quad (10)$$

Each computational stage is carried out by using 2-point FT, but readdressing is necessary between stages. The loops which implement these stages were discussed in chapter 2.

Direct computation shows that Q satisfies the condition

$$Q(\underline{x}_1 \otimes \underline{x}_2 \otimes \underline{x}_3) = \underline{x}_3 \otimes \underline{x}_2 \otimes \underline{x}_1, \quad (11)$$

where $\underline{x}_1, \underline{x}_2, \underline{x}_3$ are 2-dimensional vectors. Since these tensor products span $\mathbf{C}^8$, condition (11) uniquely defines Q. We call Q the *8-point bit-reversal* for the following reason. Each integer $0 \le n < 8$ can be uniquely written as

$$n = a_0 + 2a_1 + 4a_2, \quad 0 \le a_0, a_1, a_2 < 2, \quad (12)$$

and we call the ordered triple,

$$(a_0, \quad a_1, \quad a_2), \quad (13)$$

the binary bit representation of n. Consider the permutation of the indexing set,

$$\Pi(a_0, a_1, a_2) = (a_2, a_1, a_0), \quad 0 \le a_0, a_1, a_2 < 2, \quad (14)$$

given by reversing the bits. The following table describes Π.

Bit-Reversal Π:

000	000
001	100
010	010
011	110
100	001
101	101
110	011
111	111

Direct computation shows that the permutation matrix correspond-ing to Π is Q.

More generally, if $N = 2^k$, the *N-point bit-reversal* is the per-mutation Q uniquely defined by the condition,

$$Q(\underline{x}_1 \otimes \ldots \otimes \underline{x}_s) = \underline{x}_s \otimes \ldots \otimes \underline{x}_1, \qquad (15)$$

where x_l is a 2-dimensional vector. Arguing as above, Q corresponds to the indexing set permutation Π given by bit-reversal. Explicitly, each integer $0 \le n < N$ can be uniquely written as

$$n = a_0 + 2a_1 + \ldots + 2^{k-1}a_{k-1}, \quad 0 \le a_n < 2, \qquad (16)$$

and we call the ordered k-tuple

$$(a_0, a_1, \ldots, a_{k-1}), \qquad (17)$$

the *binary bit representation* of n. Define the permutation

$$\Pi(a_0, a_1, \ldots, a_{k-1}) = (a_{k-1}, \ldots, a_1, a_0). \qquad (18)$$

The corresponding permutation matrix satisfies (15) and is N-point bit-reversal.

Denote N-point bit reversal by $Q(N)$, $N = 2^k$. We will show that

$$Q(N) = P(N)(I_2 \otimes P(N/2)) \ldots (I_{N/4} \otimes P(4)). \qquad (19)$$

Define the sequence of permutations

$$Q_1 = P(n), \tag{20}$$

$$Q_2 = I_2 \otimes P(N/2), \tag{21}$$

$$\vdots$$

$$Q_{k-1} = I_{N/4} \otimes P(4) = I_2 \otimes (I_{N/8} \otimes P(4)). \tag{22}$$

By (4),

$$Q_2 \ldots Q_{k-1} = I_2 \otimes Q', \tag{23}$$

where

$$Q' = P(N/2)(I_2 \otimes P(N/4)) \ldots (I_{N/8} \otimes P(4)) \tag{24}$$

which we see is the factorization on the right-hand side of (19) corresponding to $N/2 = 2^{k-1}$. By induction, we can assume $Q' = Q(N/2)$ is $N/2$-point bit-reversal. Definition (15) implies

$$(Q_2 \ldots Q_{k-1})(\underline{x}_1 \otimes \underline{x}_2 \ldots \otimes \underline{x}_k) = \underline{x}_1 \otimes (\underline{x}_k \otimes \ldots \otimes \underline{x}_2). \tag{25}$$

We complete the inducction step using

$$P(N)(\underline{x}_1 \otimes \underline{x}) = \underline{x} \otimes \underline{x}_1, \tag{26}$$

where $\underline{x}$ is a $N/2$-dimensional vector.

In the same way, we can show that

$$Q(N) = (P(4) \otimes I_{N/4}) \ldots (P(N/2) \otimes I_2)P(N). \tag{27}$$

By induction on transform size, we have the following result of *Gentleman-Sande* [4].

Theorem 1. If $N = 2^k$ then

$$F(N) = Q(N) \prod_{l=1}^{k} (I_{2^{l-1}} \otimes T(2^{k-l+1}))(I_{2^{l-1}} \otimes F(2) \otimes I_{2^{k-l}}), \tag{28}$$

where the product is read right to left.

By this factorization, $F(N)$ is decomposed into k computational stages.

$$X_1 = (T_{2^{k-1}}(2^k))(F(2) \otimes I_{2^{k-1}}), \tag{29}$$

$$X_2 = (I_2 \otimes T_{2^{k-2}}(2^{k-1}))(I_2 \otimes F(2) \otimes I_{2^{k-2}}), \tag{30}$$

$$\vdots,$$

$$X_k = I_{2^{k-1}} \otimes F(2). \tag{31}$$

In general,

$$X_l = (I_{2^{l-1}} \otimes T_{2^{k-l}}(2^{k-l+1}))(I_{2^{l-1}} \otimes F(2) \otimes I_{2^{k-l}}). \tag{32}$$

In the l-th stage, we have 2^{l-1} copies of the 2-point FT on the vectors of size 2^{k-l} followed by 2^{l-1} copies of the twiddle factor $T_{2^{k-l}}(2^{k-l+1})$. It can be viewed as a parallel action on vectors of size 2^{k-l}. Vector length varies through the computation from 2^{k-1} to 1.

Taking the transpose yields the *Cooley-Tukey radix-2 FFT algorithm* [3].

Theorem 2. If $N = 2^k$ then

$$F(N) = \prod_{l=1}^{k} (I_{2^{k-l}} \otimes F(2) \otimes I_{2^{l-1}})(I_{2^{k-l}} \otimes T_{2^{l-1}}(2^l)))Q(N).$$

The Cooley-Tukey FFT has bit-reversal at input (*decimation -in-time*), while the Gentleman-Sande FFT has bit-reversal at output (*decimation-in-frequency*). As written, Gentleman-Sande FFT performs an FFT followed by a twiddle factor at every stage, but regrouping reverses the order. It is standard to combine these steps in code.

4.3. Pease FFT Algorithm

In [6], Pease designed a variation of the Cooley-Tukey FFT which he asserts is 'better adapted to parallel processing in a special purpose machine'. A few examples will show what he had in mind.

Set $P(N) = P(N, M)$ and $T(N) = T_M(N)$ with $N = 2^k$ and $M = 2^{k-1}$.

Example 1. Consider the factorization

$$F(4) = P(4)(I_2 \otimes F(2))T(4)(F(2) \otimes I_2). \qquad (1)$$

In section 2, we described the data flow of the corresponding computation.

One of the main features of the Pease FFT is that we have constant data flow in all stages of the computation. To accomplish this, we use the commutation theorem in the form

$$F(2) \otimes I_2 = P(4)(I_2 \otimes F(2))P(4). \qquad (2)$$

Placing (2) in (1), we have

$$F(4) = P(4)(I_2 \otimes F(2))T(4)P(4)(I_2 \otimes F(2))P(4). \qquad (3)$$

The smaller size FFT factors are all the same. This factor, as discussed in chapter 2, is especially suited for parallel processing. The data flow is now explicitly part of the factorization and is constant throughout the the computation. As envisioned by Pease, a single hardwired device can implement the action of $P(4)$.

Example 2. We will derive a variation of the factorization (2.6) where each stage of the computation has same data flow. Set

$$P = P(8, 2). \qquad (4)$$

From chapter 2,

$$P^2 = P(8, 4) = P^{-1}, \qquad (5)$$

$$P^3 = I_8. \qquad (6)$$

By the commutation theorem,

$$F(2) \otimes I_4 = P(I_4 \otimes F(2))P^{-1}, \tag{7}$$

$$I_2 \otimes F(2) \otimes I_2 = P^2(I_4 \otimes F(2))P^{-2}. \tag{8}$$

Placing these identities in the factorization in theorem (2.1),

$$F(8) = Q(8)(I_4 \otimes F(2))(I_2 \otimes T(4))$$

$$(I_2 \otimes F(2) \otimes I_2)T(8)(F(2) \otimes I_4), \tag{9}$$

we get

$$F(8) = Q(8)(I_4 \otimes F(2))(T_2 \otimes T(4))$$

$$P^2(T_4 \otimes F(2))P^{-2}T(8)P(I_4 \otimes F(2))P^{-1}. \tag{10}$$

Diagonal matrices remain diagonal matrices upon conjugating by permutation matrices. This idea will be used repeatedly to change data flow at the cost of changing twiddle factors. Setting

$$T_2 = P(I_2 \otimes T(4))P^{-1} = I_2 \otimes T(4), \tag{11}$$

$$T_1 = P^{-1}T(8)P = T_2(8), \tag{12}$$

we can rewrite (10) as

$$F(8) = Q(8)(I_4 \otimes F(2))P^{-1}T_2(I_4 \otimes F(2)P^{-1}T_1(I_4 \otimes F(2))P^{-1}. \tag{13}$$

Since $P^{-1} = P(8,4)$, we have *Pease 8-point FT*.

   Theorem 1. If $N = 8$,

$$F(8) = Q(8)(I_4 \otimes F(2))P(8,4)T_2(I_4 \otimes F(2))$$

$$P(8,4)T_1(I_4 \otimes F(2))P(8,4), \tag{14}$$

where $T_1 = T_2(8)$ and $T_2 = I_2 \otimes T_2(4)$.

We distinguish three computational stages,

$$X_l = T_l(I_4 \otimes F(2))P(8,4), \quad l = 1,2,3, \; T_3 = I_8, \qquad (15)$$

which differ only in the twiddle factors.

To derive the general case, we consider factorization (2.29). The goal is to design an algorithm having the same data flow in each stage of the computation. Set

$$P = P(N,2), \quad N = 2^k. \qquad (16)$$

In chapter 2, we proved that

$$P^l = P(N,2^l), \quad P^k = I_N. \qquad (17)$$

Using the identity,

$$I_{2^{l-1}} \otimes F(2) \otimes I_{2^{k-l}} = P^l(I_{2^{k-1}} \otimes F(2))P^{-l}, \qquad (18)$$

in the Gentleman-Sande FT algorithm, we get

$$F(N) = Q(N) \prod_{l=1}^{k} (I_{2^{l-1}} \otimes T(2^{k-l+1}))P^l(I_{2^{k-1}} \otimes F(2))P^{-l}. \qquad (19)$$

Set

$$T_l = P^{-l}(I_{2^{l-1}} \otimes T(2^{k-l+1}))P^l = P^{-1}(T(2^{k-l+1}) \otimes I_{2^{l-1}})P. \qquad (20)$$

We can rewrite (19) by regrouping the factors as in theorem 1.

$$F(N) = Q(N) \prod_{l=1}^{k} T_l(I_{2^{k-1}} \otimes F(2))P^{-1}. \qquad (21)$$

Since $P^{-1} = P(2^k, 2^{k-1})$, we have the *generalized Pease FFT*.

Theorem 2. If $N = 2^k$, then

$$F(N) = Q(N) \prod_{l=1}^{k} T_l(I_{2^{k-1}} \otimes F(2))P(2^k, 2^{k-1}), \qquad (22)$$

where T_l is the diagonal matrix defined in (20).

Peases FFT has the advantage of complete parallelization and constant data flow. The twiddle factors change in each stage.

A vectorized variation of the Pease FFT was presented by Korn-Lambiotte [5] for implementation on the STAR 100. We will design two vector variations. In each, we replace the parallel operation $I_{2^{k-1}} \otimes F(2)$ with the vector operation $F(2) \otimes I_{2^{k-1}}$. Again, set $P = P(2^k, 2)$.

Pease FFT can be vectorized by replacing $I_{2^{k-1}} \otimes F(2)$ with $P^{-1}(F(2) \otimes I_{2^{k-1}})P$ in (22) as follows.

$$F(N) = Q(N) \prod_{l=1}^{k} T_l P^{-1}(F(2) \otimes I_{2^{k-1}}). \tag{23}$$

Since

$$PT_l P^{-1} = P(2^k, 2^{k-l+1})(I_{2^{l-1}} \otimes T(2^{k-l+1}))P(2^k, 2^{l-1}) \tag{24}$$

$$= T(2^{k-l+1}) \otimes I_{2^{l-1}}, \tag{25}$$

we have the *Korn-Lambiotte FFT*.

Theorem 3. If $N = 2^k$ then

$$F(N)$$

$$= Q(N) \prod_{l=1}^{k} P(2^k, 2^{k-1})(T(2^{k-l+1}) \otimes I_{2^{l-1}})(F(2) \otimes I_{2^{k-1}}). \tag{26}$$

Korn-Lambiotte FFT has complete vectorization and constant data flow. Only the twiddle factor varies at different stages of the computation.

Factorization (22) and (26) are decimation-in-frequency since the output is bit-reversed. Taking transpose results in decimation-in-time, since now the input is bit-reversed. This form is due to *Singleton* [7].

4.4. Auto-sorting FFT Algorithm

The cost of performing N-point bit-reversal either on output or input data can be an important part of the overall cost of an FFT computation on many machines. In *Cochran*, et al. [2], an FFT algorithm, attributed to Stockham, is designed which computes the FFT in proper order without requiring permutation either after or before computational stages. We call such an algorithm an *auto-sort algorithm*. *Temperton* examines, in detail, the implementation of the Stockham FFT and mixed radix generalizations on the CRAY-1 in a series of papers [10, 11, 12].

The main idea underlying the Stockham auto-sort FFT is to distribute the N-point bit-reversal throughout the different stages of the computation. At the same time, all the FFT computations are unchanged. However, there is a trade-off. First, the data flow in each stage of the Pease FFT is the same, where in the Stockham FFT the data flow varies from stage to stage. Also, the data permutation required in each of the computational stages of the Pease FFT can be effectively implemented using generally available features of vector machines. In particular, in the radix-2 case, the perfect shuffle

$$P^{-1} = P(N, N/2), \quad N = 2^k \tag{1}$$

is matched to the vector construction

$$\text{stride by } N/2 \tag{2}$$

applied to the N-point data. In the Stockham FFT, the corresponding data permutations can also be implemented using the vector instruction 'stride' but it operates on data of varying sizes analogous to the changing data flow or vector lengths in the Cooley-Tukey FFT.

Example 1. Denote 4-point bit-reversal by $Q(4)$ and 8-point bit-reversal by $Q(8)$. Then

$$Q(8)(I_4 \otimes F(2))Q(8)^{-1} = F(2) \otimes I_4, \tag{3}$$

$$(Q(4) \otimes I_2)(I_2 \otimes F(2) \otimes I_2)(Q(4)^{-1} \otimes I_2) \;=\; F(2) \otimes I_4. \quad (4)$$

To derive the 8-point FFT, first vectorize the factorization (3.4), using bit-reversals :

$$F(8) = (F(2) \otimes I_4)Q(8)(I_2 \otimes T(4))(Q(4)^{-1} \otimes I_2)$$

$$(F(2) \otimes I_4)(Q(4) \otimes I_2)T(8)(F(2) \otimes I_4). \quad (5)$$

Again, we conjugate the twiddle factors by permutations and obtain

$$F(8) = (F(2) \otimes I_4)Q(8)(Q(4)^{-1} \otimes I_2)T_2$$

$$(F(2) \otimes I_4)(Q(4_\otimes I_2)T(8)(F(2) \otimes I_4), \quad (6)$$

where $T_2 = (Q(4) \otimes I_2)(I_2 \otimes T(4))(Q(4)^{-1} \otimes I_2)$. Since

$$Q(8)(Q(4) \otimes I_2) = P(8,2), \quad (7)$$

$$Q(4) = P(4,2), \quad (8)$$

we have the *8-point Stockham FFT*.

Theorem 1. If $N = 8$ then

$$F(8) = (F(2) \otimes I_4)P(8,2)T_2(F(2) \otimes I_4)$$

$$(P(4,2) \otimes I_2)T_1(F(2) \otimes I_4), \quad (9)$$

where $T_1 = T_4(8)$ and $T_2 = (P(4,2) \otimes I_2)(I_2 \otimes T_2(4))(P(4,2) \otimes I_2)$.

The Stockham FFT is a completely vectorized FFT in which bit-reversal has been distributed through the computation. Data flow is no longer constant between stages.

To derive the general Stockham FFT, we use

$$Q_l = Q(2^l) \otimes I_{2^{k-l}}. \quad (10)$$

Placing the identity

$$I_{2^{l-1}} \otimes F(2) \otimes I_{2^{k-l}} = Q_l^{-1}(F(2) \otimes I_{2^{k-1}})Q_l, \qquad (11)$$

in the Gentleman-Sande FFT, we get

$$F(N) = Q_k \prod_{l=1}^{k}(I_{2^{l-1}} \otimes T(2^{k-l+1}))Q_l^{-1}(F(2) \otimes I_{2^{k-1}})Q_l \qquad (12)$$

$$= Q_k \prod_{l=1}^{k} Q_l^{-1}T_l(F(2) \otimes I_{2^{k-1}})Q_l, \qquad (13)$$

where $T_l = Q_l(I_{2^{l-1}} \otimes T(2^{k-l+1}))Q_l^{-1}$. Regrouping, we can write

$$F(N) = Q_k Q_k^{-1} \prod_{l=1}^{k} T_l(F(2) \otimes I_{2^{k-1}})Q_l Q_{l-1}^{-1}, \qquad (14)$$

with $Q_0 = I_{2^k}$. Since $Q_l Q_{l-1}^{-1} = P(2^l, 2) \otimes I_{2^{k-l}}$, we have proved the *Stockham FFT*.

Theorem 2. If $N = 2^k$ then

$$F(N) = \prod_{l=1}^{k} T_l(F(2) \otimes I_{2^{k-1}})(P(2^l, 2) \otimes I_{2^{k-l}}), \qquad (15)$$

where T_l is the diagonal matrix

$$T_l = (Q(2^l) \otimes I_{2^{k-l}})(I_{2^{l-1}} \otimes T_{2^{k-l}}(2^{k-l+1}))(Q(2^l) \otimes I_{2^{k-l}}). \qquad (16)$$

4.5. Mixed Radix Cooley-Tukey FFT Algorithm

Historically, radix-2 algorithms were the first to be designed and dominated on serial machines. On vector machines, where arithmetic processing is very fast, the cost of data transfer becomes a significantly more important ratio of the overall cost. Radix-4 algorithms reduce this data transfer cost without appreciably increasing arithmetic cost. Agarwal-Cooley have designed radix-4 FFT algorithms for implementation on the IBM 3090 vector facility. Mixed radix FFT offer additional tools for utilizing high-speed processing without being hampered by data transfer problems. The theory underlying mixed radix FFT algorithms has been developed in several papers and is completely analogous to the theory developed in the preceeding sections. (See [8] for an early account.) In this section, we generalize the radix-2 Cooley-Tukey FFT to the mixed radix case.

The mixed radix Gentleman-Sande FFT algorithm is based on the factorization

$$F(N) = P(N,S)(I_R \otimes F(S))T_S(N)(F(R) \otimes I_S), \quad N = RS. \quad (1)$$

Suppose $N = N_1 N_2 N_3$. By (1), with $R = N_2$ and $S = N_3$, we have

$$F(N_2 N_3) = P(N_2 N_3, N_3)(I_{N_2} \otimes F(N_3))T_{N_3}(N)(F(N) \otimes I_{N_3}). \quad (2)$$

Taking $R = N_1$ and $S = N_2 N_3$, we have

$$F(N) = P(N, N_2 N_3)(I_{N_1} \otimes F(N_2 N_3))$$

$$T_{N_2 N_3}(N)(F(N_1) \otimes I_{N_2 N_3}). \quad (3)$$

Placing (2) into (3) and using the general tensor product identity

$$I \otimes (BC) = (I \otimes B)(I \otimes C), \quad (4)$$

we have the *three-factor mixed radix FFT*.

Theorem 1. If $N = N_1 N_2 N_3$, then

$$F(N) = Q(I_{N_1 N_2} \otimes F(N_3))$$

$$T_2(I_{N_1} \otimes F(N_2) \otimes I_{N_3})T_1(F(N_1) \otimes I_{N_2 N_3}), \quad (5)$$

where Q is the permutation matrix

$$Q = P(N, N_2 N_3)(I_{N_1} \otimes P(N_2 N_3, N_3)) \quad (6)$$

and T_1 and T_2 are the diagonal matrices

$$T_1 = T_{N_2 N_3}(T), \tag{7}$$

$$T_2 = I_{N_1} \otimes T_{N_3}(N_2 N_3). \tag{8}$$

Viewed as a vector computation, the operation $F(N_1) \otimes I_{N_2 N_3}$ is the vector N_1-point FFT on vectors of length $N_2 N_3$, the operation $I_{N_1} \otimes F(N_2) \otimes I_{N_3}$ is N_1 independent vector N_2-point FFT on vector of length N_3, and the operation $I_{N_1 N_2} \otimes F(N_3)$ is $N_1 N_2$ independent vector N_3-point FFT's. In particular, vector length varies as follows.

$$N_2 N_3, \quad N_3, \quad 1. \tag{9}$$

The *general mixed radix factorization* is proved by induction on the number of factors. Suppose

$$N = N_1 N_2 \ldots N_k, \tag{10}$$

$$M_l = N_l \ldots N_k. \tag{11}$$

We see that

$$N/M_1 = 1, \quad N/M_2 = N_1, \quad \ldots, \quad N/M_k = N_1 \ldots N_{k-1} \tag{12}$$

__Theorem__ 2. If $N = N_1 N_2 \cdots N_k$ then

$$F(N) = Q \prod_{l=1}^{k} (I_{N/M_l} \otimes T_{M_{l+1}}(M_l))(I_{N/M_l} \otimes F(N_l) \otimes I_{M_{l+1}}). \tag{13}$$

where the output permutation Q is the mixed radix analog of bit-reversal. It is given by the factorization

$$Q = P(N, M_2)(I_{N/M_2} \otimes P(M_2, M_3)) \cdots$$

$$(I_{N/M_{k-1}} \otimes P(M_{k-1}, M_k)). \tag{14}$$

Direct computation shows that

$$Q(\underline{x}_1 \otimes \cdots \otimes \underline{x}_s) = \underline{x}_s \otimes \cdots \otimes \underline{x}_1, \quad \underline{x}_l \in \mathbf{C}^{N_l}. \qquad (15)$$

Formula (19) serves to define Q uniquely. We call Q, the *N-point data transposition* corresponding to the 'ordered' factorization (10).

As with bit-reversal, Q is built from smaller size data transpositions. Let Q' be N/N_1-point data transposition corresponding to the order factorization (10). Then

$$Q = P(N, M_2)(I_{N_1} \otimes Q'). \qquad (16)$$

Factorization (13) leads to a decimation-in-frequency algorithm. The corresponding decimation-in-time algorithm is derived by taking transpose on both sides of (13). We state the result in the following theorem.

Theorem 3. If $N = N_1 N_2 \cdots N_k$ then

$$F(N) = (\prod_{l=1}^{k}(I_{N/M_{k-l+1}} \otimes F(N_{k-l+1}) \otimes I_{M_{k-l+2}})$$

$$(I_{N/M_{k-l+1}} \otimes T_{M_{k-l+2}}(M_{k-l+1})))Q. \qquad (17)$$

4.6. Mixed Radix Agarwal-Cooley FFT Algorithm

A generalization of the radix-2 Pease FFT to mixed radix was designed by *Agarwal-Cooley* [1] for implementation on the IBM 3090 Vector computer. The goal, as stated, is to produce a fully vectorized mixed radix FFT algorithm requiring all the loads/stores with only small strides.

Consider $N = N_1 N_2 N_3$. By the transpose of the factorization given in theorem (5.1), we have

$$F(N) = (F(N_1) \otimes I_{N_1 N_2})T_1(I_{N_1} \otimes F(N_2) \otimes I_{N_3})$$

$$T_2(I_{N_1 N_2} \otimes F(N_3))Q, \tag{1}$$

where $T_1 = T_{N_2 N_3}(N)$ and $T_2 = I_1 \otimes (N_2 N_3)$. Set $P_l = P(N, N_l)$. By the commutation theorem, we can rewrite (1) as

$$F(N) = (F(N_1) \otimes I_{N_2 N_3})T_1 P_1(F(N_2) \otimes I_{N_1 N_3})(P_1^{-1} T_2 P_1)$$

$$(P_1^{-1} P_3^{-1})(F(N_3) \otimes I_{N_1 N_2})P_3 Q. \tag{3}$$

Applying the commutation theorem once more,

$$P_1^{-1} T_2 P_1 = P_1^{-1}(I_{N_1} \otimes T_{N_3}(N_2 N_3))P_1 = T_{N_3}(N_2 N_3) \otimes I_{N_1}. \tag{4}$$

Since $P_2 = P_1^{-1} P_3^{-1}$, we have proved the *three factor Agarwal-Cooley algorithm*.

Theorem 1. If $N = N_1 N_2 N_3$, then

$$F(N) = (F(N_1) \otimes I_{N_2 N_3})T_1' P_1(F(N_2) \otimes I_{N_1 N_3})$$

$$T_2' P_2(F(N_3) \otimes I_{N_1 N_2})P_3 Q, \tag{5}$$

where $T_1' = T_{N_1 N_2}(N)$, $T_2' = T_{N_3}(N_2 N_3) \otimes I_{N_1}$ and $P_l = P(N, N_l)$, $1 \leq l \leq 3$.

This factorization offers complete vectorization and data flow given by small strides $P(N, N_l)$, $l = 1, 2, 3$.

To prove the general case, $N = N_1 N_2 \ldots N_k$, first vectorize (5.17)

$$F(N) = (\prod_{l=1}^{k} P(N, M_{k-l+1})^{-1}(F(N_{k-l+1}) \otimes I_{N/N_{k_l+1}})$$

$$P(N, M_{k-l+1})T_l)Q. \tag{6}$$

Regrouping and using the identity,

$$P(N, M_{k-l})P(N, M_{k-l+1})^{-1} = P(N, N_{k-l}), \tag{7}$$

we have the *general Agarwal-Cooley FFT.*

Theorem 2. If $N = N_1 N_2 \cdots N_k$ then

$$F(N) = (\prod_{l=1}^{k}(F(N_{k-l+1}) \otimes I_{N/N_{k_l+1}})T_l' P(N, N_{k-l+1}))Q, \qquad (8)$$

where

$$T_l' = T_{M_{k-l}}(M_{k-l+1}) \otimes I_{N/M_{k-l+1}}. \qquad (9)$$

In the special case, $N_l = 2, l = 1, 2, \cdots, k$, we have the Korn-Lambiotte FFT.

4.7. Mixed Radix Auto-sorting FFT Algorithm

Consider $N = N_1 N_2 N_3$ and the factorization of theorem (5.1).

$$F(N) = Q(I_{N_1 N_2} \otimes F(N_3))$$

$$T_2(I_{N_1} \otimes F(N_2) \otimes I_{N_3})T_1(F(N_1) \otimes I_{N_2 N_3}). \qquad (1)$$

Q is data transposition corresponding to the ordered factorization

$$N = N_1 N_2 N_3. \qquad (2)$$

It follows that

$$Q(I_{N_1 N_2} \otimes F(N_3)) = (F(N_3) \otimes I_{N_1 N_2})Q. \qquad (3)$$

Let Q_2 denote data transposition corresponding to the ordered factorization

$$N/N_3 = N_1 N_2, \qquad (4)$$

and set

$$R_2 = Q_2 \otimes I_{N_3}. \qquad (5)$$

Observe that

$$Q_2 = P(N/N_3, N_2).$$

Since

$$R_2(I_{N_1} \otimes F(N_2) \otimes I_{N_3})R_2^{-1} = F(N_2) \otimes I_{N/N_2}, \qquad (7)$$

we can rewrite (1) as

$$F(N) = ((F(N_3) \otimes I_{N_1 N_2})QR_2^{-1})$$

$$(T_2'(F(N_2) \otimes I_{N_1 N_3})R_2)(T_1(F(N_1) \otimes I_{N_2 N_3})) \qquad (8)$$

where T_2' is the diagonal matrix

$$T_2' = R_2 T_2 R_2^{-1}. \qquad (9)$$

Direct computation shows that

$$QR_2^{-1} = P(N, N_3) \qquad (10)$$

proving the *three factor mixed radix auto-sorting FFT algorithm.*

Theorem 1. If $N = N_1 N_2 N_3$ then

$$F(N) = (F(N_3) \otimes I_{N_1 N_2})P(N, N_3)T_2'(F(N_2) \otimes I_{N_1 N_3})$$

$$(P(N_1 N_2, N_2) \otimes I_{N_3})T_1'(F(N_1) \otimes I_{N_2 N_3}), \qquad (11)$$

where $T_1' = T_1 = T_{N_2 N_3}(N)$ and

$$T_2' = (Q_2 \otimes I_{N_3})(I_{N_1} \otimes T_{N_3}(N_2 N_3))(Q_2^{-1} \otimes I_{N_3}). \qquad (12)$$

The general mixed radix auto-sort FFT is derived using the same arguments. To be precise, denote by Q_l data transposition corresponding to the ordered factorization

$$N/M_{l+1} = N_1 \ldots N_l, \qquad (13)$$

and set

$$R_l = Q_l \otimes I_{M_{l+1}}, \quad 2 \leq l \leq k. \qquad (14)$$

In particular, $R_k = Q$. The identity

$$R_l(I_{N/M_l} \otimes F(N_l) \otimes I_{M_{l+1}})R_l^{-1} = F(N_l) \otimes I_{N/N_l} \qquad (15)$$

placed into theorem (5.2) gives

$$F(N) = Q \prod_{l=1}^{k} (I_{N/M_l} \otimes T(M_l))R_l^{-1}(F(N_l) \otimes I_{N/N_l})R_l. \qquad (16)$$

Set

$$T_l' = R_l(I_{N/M_l} \otimes T(M_l))R_l^{-1}. \qquad (17)$$

Regrouping the factors in (16), we get

$$F(N) = QR_k^{-1} \prod_{l=1}^{k} T_l'(F(N_l) \otimes I_{N/N_l})R_l R_{l-1}^{-1}. \qquad (18)$$

Since we have the identity,

$$R_l R_{l-1}^{-1} = Q_l(Q_{l-1}^{-1} \otimes I_{N_l}) \otimes I_{M_{l+1}} = P(N/M_{l+1}, N_l), \qquad (19)$$

the mixed radix auto-sort FFT follows.

Theorem 2. If $N = N_1 \dots N_k$ then

$$F(N) = \prod_{l=1}^{k} T_l'(F(N_l) \otimes I_{N/N_l})P(N/M_{l+1}, N_l), \qquad (20)$$

where T_l' is the diagonal matirx

$$T_l' = (Q_l \otimes I_{M_{l+1}})(I_{N/M_l} \otimes T_{M_{l+1}}(M_l))(Q_l \otimes I_{M_{l+1}}), \qquad (21)$$

with Q_l data transposition relative to the factorization $N/M_{l+1} = N_1 \dots N_l$.

Summary

I. $N = 2^k$

$$T(2^{l+1}) = T_{2^l}(2^{l+1}).$$

$$P(2^{l+1}) = P(2^{l+1}, 2^l).$$

$Q(2^k)(a_1 \otimes \ldots \otimes a_k) = a_k, a_{k-1} \otimes \ldots a_1$ (bit reversal).

Gentleman-Sande

$$F(N) = Q(N) \prod_{l+1}^{k} (I_{2^{l-1}} \otimes T(2^{k-l+1}))(I_{2^l-1} \otimes F(2) \otimes I_{2^{k-l}}).$$

Cooley-Tukey

$$F(N) = (\prod_{l=1}^{k} I_{2^{k-l}} \otimes F(2) \otimes I_{2^{l-1}})(I_{2^{k-l}} \otimes T(2^l)))Q(N)$$

Pease

$$F(N) = Q(N) \prod_{l=1}^{k} T_l (I_{2^{k-1}} \otimes F(2)) P(2^k, 2^{k-1}),$$

$$T_l = P^{-1}(T(2^{k-l+1}) \otimes I_{2^{l-1}}) P = P^{-l}(I_{2^{l-1}} \otimes T(2^{k-l+1})) P^l$$

Korn-Lambiotte

$$F(N) = Q(N) \prod_{l=1}^{k} P(2^k, 2^{k-1})(T(2^{k-l+1}) \otimes I_{2^{l-1}})(F(2) \otimes I_{2^{k-1}}).$$

Auto-Sort

$$F(N) = \prod_{l=1}^{k} T_l (F(2) \otimes I_{2^{k-1}})(P(2^l, 2) \otimes I_{2^{k-l}}),$$

$$T_l = (Q(2^l) \otimes I_{2^{k-l}})(I_{2^{l-1}} \otimes T(2^{k-l+1}))(Q(2^l) \otimes I_{2^{k-l}}).$$

II. Mixed Radix

$$N = N_1 \ldots N_k.$$

$$M_l = N_l \ldots N_k.$$

$$T(M_l) = T_{M_{l+1}}(M_l)$$

$$Q_l(a_{N_1} \otimes \ldots \otimes a_{N_l}) = a_{N_l} \otimes \ldots \otimes a_{N_1} \text{ (bit reversal)}.$$

Gentleman-Sande

$$F(N) = Q \prod_{l=1}^{k} (I_{N/M_l} \otimes T(M_l))(I_{N/M_l} \otimes F(N_l) \otimes I_{M_{l+1}}).$$

Cooley-Tukey

$$F(N) = (\prod_{l=1}^{k} (I_{N/M_{k-l+1}} \otimes F(N_{k-l+1}) \otimes I_{N_{k-l+2}})$$

$$(I_{N/M_{k-l+1}} \otimes T(M_{k-l+1})))Q.$$

Agarwal-Cooley

$$F(N) = (\prod_{l=1}^{k} (F(N_{k-l+1}) \otimes I_{N/N_{k-l+1}})T'_l P(N, N_{k-l+1}))Q,$$

$$T'_l = T(M_{k-l+1}) \otimes I_{N/M_{k-l+1}}.$$

Auto-Sort

$$F(N) = \prod_{l=1}^{k} T'_l (F(N_l) \otimes I_{N/N_l})P(N/M_{l+1}, N_l),$$

$$T'_l = (Q_l \otimes I_{M_{l+1}})(I_{N/M_l} \otimes T(M_l))(Q_l \otimes I_{M_{l+1}}).$$

[References]

[1] Agarwal R. C. and Cooley, J. W. "Vectorized Mixed Radix Discrete Fourier Transform Algorithms". March 1986, IBM Report.

[2] Cochran, W.T. et al. "What is the Fast Fourier Transform?," IEEE Trans. Audio Electroacoust., vol.15, 1967, pp.45-55.

[3] Cooley, J.W. and Tukey, J.W. "An ALgorithm for the Machine Calculation of Complex Fourier Series," Math. Comp., vol.19, 1965, pp.297-301.

[4] Gentleman, W.M. and Sande, G. "Fast Fourier Transform for Fun and Profit," Proc. AFIPS, Joint Computer Conference, vol29, 1966, pp.563-578.

[5] Korn, D.G., Lambiotte, J.J. "Computing the Fast Fourier Transform on a Vector Computer," Math. Comp., vol.33, 1979, pp.977-992.

[6] Pease, M. C. "An Adaptation of the Fast Fourier Transform for Parallel Processing," J. ACM, vol.15, 1968, pp.252-265.

[7] Singleton, R. C. "On Computing the Fast Fourier Transform," J. ACM, vol.10, 1967, pp.647-654.

[8] Singleton, R. C. " An Algorithm for Computing the Mixed Radix Fast Fourier Transform," IEEE Trans. Audio Electroacoust.,vol.17, 1969, pp.93-103.

[9] Temperton, C. "Self-Sorting Mixed Radix Fast Fourier Transforms," Journal of Computational Physics, vol.52, 1983, pp1-23.

[10] Temperton, C. "Implementation of a Prime Factor FFT algorithm on the CRAY-1," 1987, to appear in J. Comput. Phys.

[11] Temperton, C. "Implementation of a Self-Sorting In-Place Prime Factor FFT algorithm," J. Comput. Phys., 58(3), 283-299, May 1985.

[12] Temperton, C. "A Note on a Prime Factor FFT," J. Comput. Phys., 52(1), 198-204, Oct. 1983.

[13] Heideman, M.T. and Burrus, C.S. "Multiply/add Tradeoff in Length-2^n FFT Algorithms", ICASSP'85, pp. 780-783.

[14] Duhamel, P. "Implementation of "Split-Radix" FFT Algorithms for Complex, Real, and Real-Symmetric Data", IEEE Trans. on ASSP, vol.34, No. 2, April 1986.

[15] Vetterli, M. and Duhamel, P. "Split-Radix Algorithms for Length p^m DFT's", ICASSP'88, pp. 1415-1418.

Problems

1. Write a code implementing each stage X_l, $1 \leq l \leq k$, of the Gentleman-Sande algorithm.

2. Write a code implementing bit reversal.

3. For $N=$ 8, 16, 32 and 64, describe the twiddle factor T_l in the Pease algorithm.

4. Derive the general form of the twiddle factor in the Pease factorization.

5. From the Pease algorithm, design an algorithm reversing the order of permutations and the twiddle factors.

6. From the Pease algorithm, design an algorithm having bit-reversal at output.

7. Determine the general form of the twiddle factors in the Stockham factorization.

8. Describe the twiddle factors in the mixed radix Agarwal-Cooley FFT algorithm.

9. Describe the twiddle factors in the mixed radix Auto-sort FFT algorithm.

Chapter 5

GOOD-THOMAS PFA

5.1. Introduction

The additive FFT algorithms of the preceeding two chapters make no explicit use of the multiplicative structure of the indexing set. We will see how this multiplicative structure can be applied, in the case of transform size $N = RS$, where R and S are relatively prime, to design a FT algorithm, similar in structure to these additive algorithms, but no longer requiring the twiddle factor multiplication. The idea is due to Good [2] in 1958 and Thomas [8] in 1963, and the resulting algorithm is called the Good-Thomas Prime Factor algorithm (PFA).

If the transform size $N = N_1 N_2$, then one form of an additive algorithm can be expressed by the factorization

$$F(N) = (F(N_1) \otimes I_{N_2})T(I_{N_1} \otimes F(N_2))P, \tag{1}$$

where P is a permutation matrix and T is a diagonal matrix or twiddle factor. Corresponding to a decomposition of the transform size N of the form $N = RS$, where R and S are relatively prime, one form of the Good-Thomas PFA is given by the factorization

$$F(N) = Q_1(F(R) \otimes I_S)(I_R \otimes F(S))Q_2, \tag{2}$$

where Q_1 and Q_2 are permutation matrices. We can rewrite (2) as

$$F(N) = Q_1(F(R) \otimes F(S))Q_2. \tag{3}$$

An obvious advantage of (3) is that the multiplications required in the twiddle factor stage of (1) are no longer necessary. Burrus and Eschenbacher in [1] and Temperton in [4] point out that a variant of (3) can be implemented in such a way that it is simultaneously self-sorting and in-place. In the preceeding chapter, these properties served to distinguish the data flow of the different additive FFT algorithms, but in no case were both present. We will discuss some of Temperton's ideas below.

5.2. Indexing by the CRT

The main tool in the indexing of input and output data for the Good-Thomas PFA is given by the CRT. Suppose

$$N = RS, \quad (R, S) = 1. \tag{1}$$

The CRT asserts the existence of a ring isomorphism

$$\delta : Z/R \times Z/S \cong Z/N, \tag{2}$$

where $Z/R \times Z/S$ denotes ring-direct product with component-wise addition and multiplication. We will take δ to be the specific ring-isomophism given by the complete system of idempotents relative to the decomposition (1) as described in chapter 1. Explicitly, elements e_1 and e_2 in Z/N can be found such that

$$e_1 \equiv 1 \bmod R, \quad e_1 \equiv 0 \bmod S, \tag{3}$$

$$e_2 \equiv 0 \bmod R, \quad e_2 \equiv 1 \bmod S. \tag{4}$$

Then

$$e_1^2 \equiv e_1 \bmod N, \quad e_2^2 \equiv e_2 \bmod N, \tag{5}$$

$$e_1 e_2 \equiv 0 \bmod N, \tag{6}$$

$$e_1 + e_2 \equiv 1 \bmod N. \tag{7}$$

Using these properties, we can prove that the mapping

$$\delta(a_1, a_2) \equiv a_1 e_1 + a_2 e_2 \bmod N, \ 0 \le a_1 < R, \ 0 \le a_2 < S. \tag{8}$$

is a ring-isomophism of $Z/R \times Z/S$ onto Z/N.

We will use (8) to define a permutation π of Z/N. First, by (8), each $a \in Z/N$ can be written uniquely as

$$a \equiv a_1 e_1 + a_2 e_2 \bmod N, \quad 0 \le a_1 < R, \, 0 \le a_2 < S. \qquad (9)$$

Since we also have that $a \in Z/N$ can be written uniquely as

$$a = a_2 + a_1 S, \quad 0 \le a_1 < R, \, 0 \le a_2 < S, \qquad (10)$$

a permutation π of Z/N can be defined by the formula

$$\pi(a_2 + a_1 S) \equiv a_1 e_1 + a_2 e_2 \bmod N, \quad 0 \le a_1 < R, \, 0 \le a_2 < S. \ (11)$$

Order the indexing set Z/N by π:

$$0, \ e_2, \ \ldots, \ (S-1)e_2, \qquad (12)$$

$$e_1, \ e_1 + e_2, \ \ldots, \ e_1 + (S-1)e_2, \qquad (13)$$

$$\vdots$$

$$(R-1)e_1, \ (R-1)e_1 + e_2, \ \ldots, \ (R-1)e_1 + (S-1)e_2, \qquad (14)$$

and denote the corresponding permutation matrix by Q. Then the matrix

$$F_\pi = Q F(N) Q^{-1}, \qquad (15)$$

is given by

$$F_\pi = \left[v^{\pi(a)\pi(b)} \right]_{0 \le a,b < N}, \quad v = exp(2\pi i/N). \qquad (16)$$

We will now explicitly describe F_π. First an example will be considered.

5.3. An Example N=15

Take $R = 3$ and $S = 5$. The idempotents are

$$e_1 = 10, \quad e_2 = 6. \tag{1}$$

The permutation matrix π of (2.11) is given as

$$\pi = \{0, 6, 12, 3, 9; \quad 10, 1, 7, 13, 4; \quad 5, 11, 2, 8, 14\}. \tag{2}$$

We distinguish three blocks by the following notations,

$$A = (0, 6, 12, 3, 9), \tag{3}$$

$$B = (10, 1, 7, 13, 4), \tag{4}$$

$$C = (5, 11, 2, 8, 14), \tag{5}$$

Each begins with a different multiple of e_1. Corresponding to the nine Cartesian products of these blocks, we have nine submatrices of F_π. Consider first the submatrix corresponding to $A \times A$. By (2.16), this submatrix is

$$\begin{bmatrix} 1 & 1 & 1 & 1 & 1 \\ 1 & w^6 & w^{12} & w^3 & w^9 \\ 1 & w^{12} & w^9 & w^6 & w^3 \\ 1 & w^3 & w^6 & w^9 & w^{12} \\ 1 & w^9 & w^3 & w^{12} & w^6 \end{bmatrix}, \quad w = exp(2\pi i/15). \tag{6}$$

Setting $u = exp(2\pi i/5)$, we can rewrite (6) as

$$\begin{bmatrix} 1 & 1 & 1 & 1 & 1 \\ 1 & u^2 & u^4 & u & u^3 \\ 1 & u^4 & u^3 & u^2 & u \\ 1 & u & u^2 & u^3 & u^4 \\ 1 & u^3 & u & u^4 & u^2 \end{bmatrix}, \quad u = exp(2\pi i/5). \tag{7}$$

We denote (7) by F_5, and following Temperton[5] call a *rotated 5-point FT*. We can relate F_5 to the 5-point FT matrix $F(5)$,

$$F(5) = \begin{bmatrix} 1 & 1 & 1 & 1 & 1 \\ 1 & u & u^2 & u^3 & u^4 \\ 1 & u^2 & u^4 & u & u^3 \\ 1 & u^3 & u & u^4 & u^2 \\ 1 & u^4 & u^3 & u^2 & u \end{bmatrix}, \tag{8}$$

in two ways. First, if in $F(5)$, we replace u by $w^{e_2} = u^2$, $F(5)$ becomes F_5. An algorithm computing the action of $F(5)$ can be modified to compute the action of F_5 by determining the consequences of this replacement through the different stages of the algorithm. Second

$$F_5 = PF(5), \tag{9}$$

where

$$P = \begin{pmatrix} 1 & 0 & 0 & 0 & 0 \\ 0 & 0 & 1 & 0 & 0 \\ 0 & 0 & 0 & 0 & 1 \\ 0 & 1 & 0 & 0 & 0 \\ 0 & 0 & 0 & 1 & 0 \end{pmatrix}. \tag{10}$$

Since both F_5 and $F(5)$ are symmetric matrices and $P^t = P^{-1}$, taking transpose on both sides of (9) gives

$$F_5 = F(5)P^{-1}. \tag{11}$$

Consider the submatrix corresponding to the cartesian product $B \times B$. Direct computation from (2.16) shows that this submatrix is

$$\begin{bmatrix} w^{10} & w^{10} & w^{10} & w^{10} & w^{10} \\ w^{10} & w & w^7 & w^{13} & w^4 \\ w^{10} & w^7 & w^4 & w & w^{13} \\ w^{10} & w^{13} & w & w^4 & w^7 \\ w^{10} & w^4 & w^{13} & w^7 & w \end{bmatrix}. \tag{12}$$

Factoring out w^{10} from (12), we can rewrite (12) as

$$v^2 F_5, \tag{13}$$

where $v = exp(2\pi i/3) = w^5$. Continuing in this way, we have

$$F_\pi = \begin{bmatrix} F_5 & F_5 & F_5 \\ F_5 & v^2 F_5 & v F_5 \\ F_5 & v F_5 & v^2 F_5 \end{bmatrix}, \tag{14}$$

which we can rewrite as

$$F_\pi = F_3 \otimes F_5, \tag{15}$$

where

$$F_3 = \begin{pmatrix} 1 & 1 & 1 \\ 1 & v^2 & v \\ 1 & v & v^2 \end{pmatrix}. \tag{16}$$

Since

$$F_3 = P'F(3), \tag{17}$$

where

$$P' = \begin{pmatrix} 1 & 0 & 0 \\ 0 & 0 & 1 \\ 0 & 1 & 0 \end{pmatrix}, \tag{18}$$

F_3 can be formed from $F(3)$ by replacing v in $F(3)$ by $w^{e_1} = v^2$.

Putting (15) into (2.15), we have

$$F(15) = Q^{-1}(F_3 \otimes F_5)Q , \tag{19}$$

where the permutation matrix Q is given in (2). Matrix Q, although not strictly a stride permutation, has a circular structure. As seen by (2), we begin by stride-6 *mod* 15 from the index 0,

$$0, \ 6, \ 12, \ 3, \ 9, \tag{20}$$

but in the second stage, instead of beginning at the index 1, we begin at the index 10 and stride-6 *mod* 15,

$$10, \ 1, \ 7, \ 13, \ 4. \tag{21}$$

In the last sweep, we begin at 5 and stride-6 *mod* 15,

$$5, \ 11, \ 2, \ 8, \ 14. \tag{22}$$

In [5] and [6], Temperton discusses the direct implementation of rotated FT's.

We can use (9) and (17) to rewrite (19) as

$$F(15) = Q^{-1}(P'F(3) \otimes PF(5))Q, \tag{23}$$

$$F(15) = Q^{-1}(P' \otimes P)(F(3) \otimes F(5))Q. \qquad (24)$$

Denote

$$Q' = Q^{-1}(P' \otimes P), \qquad (25)$$

we have that

$$F(15) = Q'(F(3) \otimes F(5))Q. \qquad (26)$$

From (11), additional factorizations of $F(15)$ can be provided. In all cases, after an initial input permutation, we compute the action of $F(3) \otimes F(5)$ and then permute the output to have the natural order. Generally, input and output permutations are not the same and are more complicated than those discussed above for direct implementation of (19). We notice, however, that no twiddle factor appears in (26).

5.4. Good-Thomas PFA for General Case

Returning to the permutation π of section 2, we set

$$A_0 = \{0, \, e_2, \, \ldots, \, (S-1)e_2\}, \qquad (1)$$

$$A_1 = \{e_1, \, e_1 + e_2, \, \ldots, \, e_1 + (S-1)e_2\}, \qquad (2)$$

$$\vdots$$

$$A_{R-1} = \{(R-1)e_1, \, (R-1)e_1 + e_2, \, \ldots, \, (R-1)e_1 + (S-1)e_2\}, \quad (3)$$

we can now write

$$\pi = (A_0, \, A_1, \, \ldots, \, A_{R-1}). \qquad (4)$$

Consider the submatrix of F_π corresponding to the Cartesian Product

$$A_j \times A_k, \qquad (5)$$

a typical component in this submatrix is given by forming first the product *mod N*,

$$(je_1 + le_2)(ke_1 + me_2), \quad 0 \le l, m < S, \tag{6}$$

which by (2.5)–(2.7) can be written as

$$jke_1 + lme_2, \quad 0 \le l, m < S. \tag{7}$$

Then, the (l, m) coefficient of this submatrix is

$$(w^{e_1})^{jk}(w^{e_2})^{lm}, \quad 0 \le l, m < S. \tag{8}$$

Set

$$F_S = [(w^{e_2})^{lm}]_{0 \le l, m < S}. \tag{9}$$

The submatrix corresponding to (5) is

$$[w^{e_1}]^{jk} F_S. \tag{10}$$

Continuing in this way,

$$F_\pi = F_R \otimes F_S, \tag{11}$$

where

$$F_R = [(w^{e_1})^{jk}]_{0 \le j, k < R}. \tag{12}$$

The matrices F_R and F_S are called *rotated FT's* by Temperton.

By (2.3) and (2.4), w^{e_1} is a primitive R-th root of unity and w^{e_2} is a primitive S-th root of unity. To see this, set

$$v = exp(2\pi i/R), \quad u = exp(2\pi i/S). \tag{13}$$

By (3),

$$e_1 = f_1 S , \quad (f_1, R) = 1. \tag{14}$$

Then

$$w^{e_1} = v^{f_1}, \tag{15}$$

showing that w^{e_1} is a primitive R-th root of unity. The corresponding result for w^{e_2} is proved in the same way. It follows that we can write

$$F_R = P_1 F(R), \tag{16}$$

where P_1 is a $R \times R$ permutation matrix, and

$$F_S = P_2 F(S), \tag{17}$$

where P_2 is a $S \times S$ permutation matrix. Placing (16) and (17) in (11), we have

$$F_\pi = (P_1 \otimes P_2)(F(R) \otimes F(S))Q, \tag{18}$$

and by (2.15),

$$F(N) = Q^{-1}(P_1 \otimes P_2)(F(R) \otimes F(S))Q, \tag{19}$$

It follows that to compute the action of $F(N)$, we begin with the input permutation Q, compute the action of the tensor product $F(R) \otimes F(S)$ and complete the computation by arranging the output in its natural order by the permutation $Q^{-1}(P_1 \otimes P_2)$. Other factorizations can be obtained by taking the transpose on both sides of (16),

$$F_R = F(R)P_1^{-1}, \tag{20}$$

and placing (20) in (11) rather than (16).

If modules exist, implementing the tensor product $F_R \otimes F_S$ then the data flow of the computation of $F(N)$ is given by the permutation Q. The permutation Q can be viewed as follows. We begin at the index point 0 and stride by e_2. This process continues where at each new stage we begin at the index point given by a multiple of e_1.

Since $F(R) \otimes F(S)$ can be viewed as S actions of $F(R)$ followed by R actions of $F(S)$, the arithmetic of (19) is given by the formulas

$$a(N) = Sa(R) + Ra(S), \tag{21}$$

$$m(N) = Sm(R) + Rm(S), \tag{22}$$

where algorithms computing R-point FT are taken which require $a(R)$ additions and $m(R)$ multiplications. The multiplications required by the twiddle factor in the additive algorithms are no longer necessary.

Formula (19) can be generalized to several factors. Suppose $N = n_1 n_2 = m_1 m_2 n_2$ where $n_1 = m_1 m_2$, m_1 and m_2 relatively prime. Then , we can write

$$F(n_1) = R'(F(m_1) \otimes F(m_2))R, \tag{23}$$

where R and R' are permutation matrices. Placing (23) into (19), we have

$$F(n) = R'''(F(m_1) \otimes F(m_2) \otimes F(n_2))R'', \tag{24}$$

where R'' and R''' are permutation matrices.

In subsequent chapters, we will combine the Good-Thomas PFA with multiplicative FT algorithms to produce several FT algorithms having distinct data flow and arithmetic. (See [3].)

5.5. Self-sorting PFA

Burrus and Eschenbacher [1] point out that Good-Thomas PFA can be computed in-place and in-order. Temperton [4,5,6] discusses the implementation of PFA's on different computer architectures, especially on CRAY. He showes that the indexing required for PFA was actually simpler than that for conventional Cooley- Tukey FFT algorithm.

Temperton implemented the $N = RS$-point FT using directly F_R and F_S. The indexing for input and output data in this case are the same.

Consider the following example. Let $N = 42 = 6 \times 7$, the corresponding system of idempotents is $\{e_1, e_2\} = \{7, 36\}$. The mapping given in (4.1)-(4.3) can be described by the 2-dimensional array,

$$\begin{bmatrix} 0 & 7 & 14 & 21 & 28 & 35 \\ 36 & 1 & 8 & 15 & 22 & 29 \\ 30 & 37 & 2 & 9 & 16 & 23 \\ 24 & 31 & 38 & 3 & 10 & 17 \\ 18 & 25 & 32 & 39 & 4 & 11 \\ 12 & 19 & 26 & 33 & 40 & 5 \\ 6 & 13 & 20 & 27 & 34 & 41 \end{bmatrix}, \tag{1}$$

We can implement this by the simple code,

```
INTEGER I(R)
DATA I/0,7,14,21,28,35/
```

Updating the indexing for each subsequent transform is achieved by simple auto-increment addressing mode,

```
       J=I(R)+1
       DO 100 K=R, 2
       I(K)=I(K-1)+1
100 CONTINUE
       I(1)=J.
```

The code requires no IF statements or address computation *mod N*. Temperton [5] describes in detail the minimum-add rotated discrete Fourier transform modules for sizes 2, 3, 4, 5, 7, 8, 9 and 16.

[References]

[1] Burrus, C.S. and Eschenbacher, P.W. "An In-place In-order Prime Factor FFT Algorithm", *IEEE Trans.*, *ASSP 29*, (1981), pp. 806-817.

[2] Good, I.J. "The Interaction Algorithm and Practical Fourier Analysis", *J. Royal Statist. 'oc., Ser.* B20 (1958):361-375.

[3] Kolba, D.P. and Parks, T.W. "A Prime Factor FFT Algorithm Using high-speed Convolution", *IEEE Trans.* ASSP 25(1977).

[4] Temperton, C. "A Note on Prime Factor FFT Algorithms", *J. Comput. Physics.*, 52 (1983), PP. 198-204.

[5] Temperton, C. "A New Set of Minimum-add Small-n Rotated DFT Modules", to appear in *J. Comput. Physics*.

[6] Temperton, C. "Implementation of A Prime Factor FFT Algorithm on CRAY-1", to appear in *Parallel Computing*.

[7] Temperton, C. "A Self-sorting In-place Prime Factor Real/half -complex FFT Algorithm", to appear in *J. Comput. Phys*.

[8] Thomas, L.H. "Using a Computer to Solve Problems in Physics", in Applications of Digital Computers, Ginn and Co., Boston, Mass., 1963.

[9] Chu, S. and Burrus, C.S. "A Prime Factor FFT Algorithm Using Distributed Arithmetic", IEEE Trans. on ASSP, Vol. 30, No. 2, pp. 217-227, April 1982.

Problems

1. Find the system of idempotents of $N = 2 \times 3$, and define the permutation matrix Q as section 2.

2. Find the system of idempotents of $N = 4 \times 5$, and define the permutation matrix Q as section 2.

3. Find F_2 and F_3 for 6-point Good-Thomas PFA based on the idempotents of Problem 1.

4. Find F_4 and F_5 for 20-point Good-Thomas PFA based on the idempotents of Problem 2.

5. Give arithmetic counts for Problem 3 and 4 by direct computation of F_2, F_3, F_4 and F_5.

6. Give arithmetic counts for 6-point and 20-point Cooley-Tukey FFT algorithm, where $F(2)$, $F(3)$, $F(4)$ and $F(5)$ are carried out by direct computation. Compare with those of Problem 5.

7. Derive a Good-Thomas PFA for $N = 75$, and give F_3 and F_{25}.

8. Derive a Good-Thomas PFA for $N = 100$, and give F_4 and F_{25}. Derive the Cooley-Tukey FFT algorithm with a factorization of $100 = 10 \times 10$. Compare the arithmetic counts of these two algorithms.

9. Give the self-sorting indexing table for $N = 40 = 5 \times 8$ as in section 5.

Chapter 6

LINEAR AND CYCLIC CONVOLUTION

Linear convolution is one of the most frequent computations carried out in digital signal processing (DSP). The standard method for computing a linear convolution is to use the convolution theorem which replaces the computation by FFT of correspondingsize. In the last ten years, theoretically better convolution algorithms have been developed. The Winograd Small Convolution algorithm [1] is the most efficient as measured by the number of multiplications.

First, we derive the *convolution theorem*, by two different methods. The second method is based on the CRT for polynomials. A special case of of the CRT is then applied in a more general setting to derive the Cook-Toom [2] algorithm. The generalized (polynomial) CRT is then used to derive the Winograd Small Convolutin algorithm. We emphasize the interplay between linear and cyclic convolution computations.

6.1. <u>Definitions</u>

Consider vectors $\underline{h}$ and $\underline{g}$ of sizes M and N, respectively. The *linear convolution* of $\underline{h}$ and $\underline{g}$ is the vector $\underline{s}$ of size $L = M + N - 1$ defined by

$$s_k = \sum_{n=0}^{k} h_{k-n} g_n, \qquad 0 \le k < L, \qquad (1)$$

where, we take $h_m = 0$ if $m \ge M$ and $g_n = 0$ if $n \ge N$.

<u>Example</u> 1. The linear convolution $\underline{s}$ of a vector $\underline{h}$ of size 2

and a vector $\underline{g}$ of size 3 is given by

$$s_0 = h_0 g_0,$$
$$s_1 = h_1 g_0 + h_0 g_1,$$
$$s_2 = h_1 g_1 + h_0 g_2,$$
$$s_3 = h_1 g_2.$$

Associate the polynomial $h(x)$ of degree $M - 1$ to the vector $\underline{h}$ of size M,

$$h(x) = h_0 + h_1 x + \cdots + h_{M-1} x^{M-1}. \tag{2}$$

Direct computation shows that formula (1) is equivalent to the polynomial product

$$s(x) = h(x) g(x). \tag{3}$$

The representation of linear convolution by polynomial product permits the application of results in polynomial rings, especially the CRT.

Example 2. Consider the linear convolution $\underline{s}$ of a vector $\underline{h}$ of size 3 and a vector $\underline{g}$ of size 4. By definition

$$s_0 = h_0 g_0,$$
$$s_1 = h_1 g_0 + h_0 g_1,$$
$$s_2 = h_2 g_0 + h_1 g_1 + h_0 g_2,$$
$$s_3 = h_2 g_1 + h_1 g_2 + h_0 g_3,$$
$$s_4 = h_2 g_2 + h_1 g_3,$$
$$s_5 = h_2 g_3.$$

The linear convolution can be described by matrix multiplication.

$$\underline{s} = \begin{bmatrix} h_0 & 0 & 0 & 0 \\ h_1 & h_0 & 0 & 0 \\ h_2 & h_1 & h_0 & 0 \\ 0 & h_2 & h_1 & h_0 \\ 0 & 0 & h_2 & h_1 \\ 0 & 0 & 0 & h_2 \end{bmatrix} \underline{g}.$$

In general, if $\underline{s}$ is the linear convolution of a vector $\underline{h}$ of size M and a vector $\underline{g}$ of size N, we can write

$$\underline{s} = H\underline{g}, \tag{4}$$

where H is the $L = M + N - 1 \times N$ matrix

$$H = \begin{bmatrix} h_0 & 0 & \cdot & \cdot & \cdot & 0 \\ h_1 & h_0 & \cdot & \cdot & \cdot & 0 \\ \cdot & \cdot & \cdot & \cdot & \cdot & \cdot \\ h_{M-1} & \cdot & \cdot & \cdot & \cdot & \cdot \\ 0 & h_{M-1} & \cdot & \cdot & \cdot & 0 \\ 0 & 0 & \cdot & \cdot & \cdot & h_0 \\ \cdot & \cdot & \cdot & \cdot & \cdot & \cdot \\ 0 & 0 & \cdot & \cdot & \cdot & h_{M-1} \end{bmatrix}. \tag{5}$$

Consider two vectors $\underline{a}$ and $\underline{b}$ of size N. The cyclic convolution $\underline{c}$ of $\underline{a}$ and $\underline{b}$, denoted by $\underline{a} * \underline{b}$ is the vector of size N defined by the formula

$$c_k = \sum_{n=0}^{N-1} a_{k-n} b_n, \qquad 0 \le n < N. \tag{6}$$

The indeces of the vectors are taken *mod N*.

Example 3. The cyclic convolution $\underline{c}$ of two vectors $\underline{a}$ and $\underline{b}$ of size 3 is given by

$$c_0 = a_0 b_0 + a_2 b_1 + a_1 b_2,$$

$$c_1 = a_1 b_0 + a_0 b_1 + a_2 b_2,$$

$$c_2 = a_2 b_0 + a_1 b_1 + a_0 b_2.$$

Observe that $a_{-1} = a_2$ and $a_{-2} = a_1$.

In Chapter 1, we discussed the quotient polynomial ring

$$\mathbb{C}\,[x]/(x^N - 1) \tag{7}$$

consisting of the set of all polynomials of degree $< N$, where addition and multiplication are taken as polynomial addition and multiplication *mod* $(x^N - 1)$.

<u>Example</u> 4. Consider two polynomials

$$a(x) = a_0 + a_1 x + a_2 x^2,$$

$$b(x) = b_0 + b_1 x + b_2 x^2.$$

The product is

$$a(x)b(x) = a_0 b_0 + (a_1 b_0 + a_0 b_1)x + (a_2 b_0 + a_1 b_1 + a_0 b_2)x^2$$

$$+(a_2 b_1 + a_1 b_2)x^3 + a_2 b_2 x^4.$$

This is the linear convolution. The product

$$c(x) \equiv a(x)b(x) \ mod \ (x^3 - 1),$$

is formed by setting $x^3 = 1$ in the expansion of the product $a(x)b(x)$. We find that the coefficients of $c(x)$ are given by

$$c_0 = a_0 b_0 + a_2 b_1 + a_1 b_2,$$

$$c_1 = a_1 b_0 + a_0 b_1 + a_2 b_2,$$

$$c_2 = a_2 b_0 + a_1 b_1 + a_0 b_2.$$

Thus multiplication in the ring

$$\mathbb{C} \ [x]/(x^3 - 1),$$

computes 3×3 cyclic convolution.

In general, multiplication in the ring (7) computes $N \times N$ cyclic convolution. To see this, consider polynomials $a(x)$ and $b(x)$ of degree $< N$, and compute the product $a(x)b(x)$,

$$a(x)b(x) = \sum_{n=0}^{2N-2} \left(\sum_{k=0}^{n} a_{n-k} b_k \right) x^n, \tag{8}$$

where $a_n = b_n = 0$ whenever $n \geq N$. Setting $x^N = 1$ in (8), we have

$$c(x) = \sum_{n=0}^{N-1} c_n x^n \equiv a(x)b(x) \bmod (x^N - 1), \tag{9}$$

where

$$c_n = \sum_{k=0}^{n} a_{n-k} b_k + \sum_{k=0}^{n+N} a_{n+N-k} b_k \tag{10}$$

$$= \sum_{k=0}^{n} a_{n-k} b_k + \sum_{k=n+1}^{N-1} a_{n+N-k} b_k \tag{11}$$

$$= \sum_{k=0}^{N-1} a_{n-k} b_k. \tag{12}$$

In (12), the indices are taken *mod N*. By the definition, we see that (9) computes $N \times N$ cyclic convolution. An important outcome of the discussion is that $N \times N$ cyclic convolution can be computed by first computing linear convolution as polynomial product, then setting $x^N = -1$.

As with linear convolution, cyclic convolution can also be expressed by matrix multiplication.

Example 5. Returning to example 3, we can write

$$\underline{c} = C\underline{b},$$

where

$$C = \begin{bmatrix} a_0 & a_2 & a_1 \\ a_1 & a_0 & a_2 \\ a_2 & a_1 & a_0 \end{bmatrix}.$$

The matrix C is an example of a circulant matrix, which we will define below. If S denotes the 3×3 cyclic shift matrix

$$S = \begin{bmatrix} 0 & 0 & 1 \\ 1 & 0 & 0 \\ 0 & 1 & 0 \end{bmatrix},$$

then

$$S^2 = \begin{bmatrix} 0 & 1 & 0 \\ 0 & 0 & 1 \\ 1 & 0 & 0 \end{bmatrix}.$$

We can write the matrix C in the form

$$C = a_0 I_3 + a_1 S + a_2 S^2.$$

The $N \times N$ *cyclic shift matrix* S is defined by the rule,

$$S\underline{x} = \begin{bmatrix} x_{N-1} \\ x_0 \\ \vdots \\ x_{N-2} \end{bmatrix}. \tag{13}$$

Observe that $S^N = I_N$. By an $N \times N$ *circulant matrix* we mean any matrix of the form

$$C = \sum_{n=0}^{N-1} a_n S^n. \tag{14}$$

At times, we will denote the dependence of C on $\underline{a}$ by writing $C(\underline{a})$.

$$C(\underline{a}) = \begin{bmatrix} a_0 & a_{N-1} & \cdot & \cdot & \cdot & a_1 \\ a_1 & a_0 & \cdot & \cdot & \cdot & a_2 \\ \cdot & a_1 & \cdot & \cdot & \cdot & \cdot \\ \cdot & & \cdot & \cdot & \cdot & \cdot \\ a_{N-1} & a_{N-2} & \cdot & \cdot & \cdot & a_0 \end{bmatrix}.$$

Example 6. The 4×4 cyclic shift matrix is

$$S = \begin{bmatrix} 0 & 0 & 0 & 1 \\ 1 & 0 & 0 & 0 \\ 0 & 1 & 0 & 0 \\ 0 & 0 & 1 & 0 \end{bmatrix}.$$

Notice that

$$S^2 = \begin{bmatrix} 0 & 0 & 1 & 0 \\ 0 & 0 & 0 & 1 \\ 1 & 0 & 0 & 0 \\ 0 & 1 & 0 & 0 \end{bmatrix}, \quad S^3 = \begin{bmatrix} 0 & 1 & 0 & 0 \\ 0 & 0 & 1 & 0 \\ 0 & 0 & 0 & 1 \\ 1 & 0 & 0 & 0 \end{bmatrix},$$

and $S^4 = I_4$. The 4×4 circulant matrix

$$C = a_0 I_4 + a_1 S + a_2 S^2 + a_3 S^3,$$

is

$$C = \begin{bmatrix} a_0 & a_3 & a_2 & a_1 \\ a_1 & a_0 & a_3 & a_2 \\ a_2 & a_1 & a_0 & a_3 \\ a_3 & a_2 & a_1 & a_0 \end{bmatrix}.$$

As we read from left to right the columns of C are cyclically shifted:

$$C = \begin{bmatrix} \underline{a} & S\underline{a} & S^2\underline{a} & S^3\underline{a} \end{bmatrix}.$$

Example 7. Denote by e_n, $0 \le n < N$, the vector of size N consisting of all zeroes except for a 1 at the n-th place. Observe that

$$e_0 * \underline{b} = \underline{b},$$

$$e_1 * \underline{b} = S\underline{b},$$

$$\vdots,$$

$$e_{N-1} * \underline{b} = S^{N-1}\underline{b}.$$

Consider the $N \times N$ cyclic convolution $\underline{c} = \underline{a} * \underline{b}$. Writing

$$\underline{a} = \sum_{n=0}^{N-1} a_n e_n, \tag{15}$$

we have

$$\underline{a} * \underline{b} = \sum_{n=0}^{N-1} a_n (e_n * \underline{b}), \tag{16}$$

which by example 7 can be rewritten as

$$\underline{a} * \underline{b} = \sum_{n=0}^{N-1} a_n S^n \underline{b}. \tag{17}$$

By (14),

$$\underline{a} * \underline{b} = C(\underline{a})\underline{b}. \tag{18}$$

The $N \times N$ cyclic convolution $\underline{c} = \underline{a} * \underline{b}$ can be computed by multiplication in the quotient polynomial ring $\mathbb{C}\,[x]/(x^N - 1)$,

$$c(x) \equiv a(x)b(x) \bmod (x^N - 1), \tag{19}$$

or by circulant matrix multiplication

$$\underline{c} = C(\underline{a})\underline{b}. \tag{20}$$

Direct computation from (20) shows that

$$C(\underline{a} * \underline{b}) = C(\underline{a})C(\underline{b}). \tag{21}$$

More generally, we can prove that the set of all $N \times N$ circulant matrices is a ring under matrix addition and multiplication and is isomorphic to the quotient polynomial ring $\mathbb{C}\,[x]/(x^N - 1)$.

6.2. Convolution Theorem

The $N \times N$ cyclic convolution can be computed using N-point FT's. This is especially convenient when efficient algorithms for N-point FT are available. The result that permits this interchange is the *Convolution Theorem*. We will give two proofs. The first depends on the representation of cyclic convolution as a matrix product by a circulant matrix. We will soon see that the FFT matrix diagonalizes every circulant matrix. The second proof uses the representation of cyclic convolution as multiplication in the quotient polynomial ring $\mathbb{C}\,[x]/(x^N - 1)$.

Example 1. Set $F = F(3)$. Denote by S the 3×3 cyclic shift matrix and by D the matrix

$$D = \begin{bmatrix} 1 & 0 & 0 \\ 0 & v & 0 \\ 0 & 0 & v^2 \end{bmatrix}, \quad v = e^{2\pi i/3}.$$

Then

$$FS = \begin{bmatrix} 1 & 1 & 1 \\ v & v^2 & 1 \\ v^2 & v & 1 \end{bmatrix} = DF,$$

which implies

$$FSF^{-1} = D.$$

D is a diagonal matrix. Thus F diagonalizes S. In addition

$$FS^2 F^{-1} = (FSF^{-1})^2 = D^2,$$

and F diagonalizes S^2.

An arbitrary 3×3 circulant matrix is of the form

$$C(\underline{a}) = a_0 I_3 + a_1 S + a_2 S^2,$$

since F diagonalizes each term of this sum, it diagonalizes $C(\underline{a})$,

$$FC(\underline{a})F^{-1} = a_0 I_3 + a_1 D + a_2 D^2.$$

Writing this out, we have

$$F \begin{bmatrix} a_0 & a_2 & a_1 \\ a_1 & a_0 & a_2 \\ a_2 & a_1 & a_0 \end{bmatrix} F^{-1} = \begin{bmatrix} g_0 & 0 & 0 \\ 0 & g_1 & 0 \\ 0 & 0 & g_2 \end{bmatrix},$$

where

$$g_0 = a_0 + a_1 + a_2,$$
$$g_1 = a_0 + v a_1 + v^2 a_2,$$
$$g_2 = a_0 + v^2 a_1 + v a_2.$$

We see that

$$FC(\underline{a})F^{-1} = diag.(\underline{g}),$$

where

$$\underline{g} = F(3)\underline{a}.$$

We can extend this argument to prove that

$$F(N)SF(N)^{-1} = D,\tag{1}$$

where S is the $N \times N$ cyclic shift matrix and D is the diagonal matrix,

$$D = \begin{bmatrix} 1 & & & & \\ & v & & & \\ & & \cdot & & \\ & & & \cdot & \\ & & & & \cdot \\ & & & & & v^{n-1} \end{bmatrix}, \quad v = exp(2\pi i/N).\tag{2}$$

It follows that

$$F(N)S^k F(N)^{-1} = D^k.\tag{3}$$

Thus

$$F(N)C(\underline{a})F(N)^{-1} = diag.(\underline{g}),\tag{4}$$

where

$$\underline{g} = F(N)\underline{a}.\tag{5}$$

In words, the N-point FFT matrix $F(N)$ diagonalizes every $N \times N$ circulant matrix $C(\underline{a})$.

Set

$$G(\underline{a}) = diag.(F\underline{a}),\tag{6}$$

where $F = F(N)$. Wwe can rewrite (4) as

$$C(\underline{a}) = F^{-1}G(\underline{a})F.\tag{7}$$

Since

$$\underline{a} * \underline{b} = C(\underline{a})\underline{b},\tag{8}$$

we have the following theorem.

Theorem 1. For vectors $\underline{a}$ and $\underline{b}$ of size N,

$$\underline{a} * \underline{b} = F^{-1}G(\underline{a})F\underline{b},\tag{9}$$

where $G(\underline{a}) = diag.(F(N)\underline{a})$.

Using (9), we can compute $\underline{a} * \underline{b}$ as follows.

1. Compute the N-point FFT of $\underline{b}$.
2. Compute the N-point FFT of $\underline{a}$.
3. Compute the diagonal matrix multiplication by $G(\underline{a})$.
4. Compute the inverse N-point FFT.

The non-symmetric role of $\underline{a}$ and $\underline{b}$ in this computation should be emphasized. In standard applications to digital filters, we fix the vector $\underline{a}$ (the elements of a linear system) then compute the cyclic convolution $\underline{a} * \underline{b}$ for many input vectors $\underline{b}$. As a consequence, the diagonal matrix $G(\underline{a})$ can be pre-computed and does not enter into the arithmetic cost of the process.

The second proof uses the CRT to 'diagonalize' multiplication in the ring $\mathbb{C}\,[x]/(x^N - 1)$. Consider the factorization

$$x^N - 1 = \prod_{n=0}^{N-1} (x - v^n), \qquad v = exp(2\pi i/N). \tag{10}$$

Applying the CRT to (10), we have that the mapping

$$a(x) \longrightarrow \begin{pmatrix} a(1) \\ a(v) \\ \vdots \\ a(v^{N-1}) \end{pmatrix} \tag{11}$$

establishes a ring-isomorphism

$$\mathbb{C}\,[x]/(x^N - 1) \cong \prod_{n=0}^{N-1} \mathbb{C}\,, \tag{12}$$

where $\prod_{n=0}^{N-1} \mathbb{C}$ denotes the ring direct product of N copies of $\mathbb{C}$ with component-wise addition and multiplication. In particular, a polynomial $a(x)$ of degree $< N$ is uniquely determined by the values $a(v^n)$, $0 \le n < N$.

Consider two polynomial $a(x)$ and $b(x)$ of degree $< N$, and set $c(x) \equiv a(x)b(x) \ mod \ (x^N - 1)$. The polynomial $c(x)$ is uniquely determined by the values $c(v^n)$, $0 \le n < N$. Since $(v^n)^N = 1$,

$$c(v^n) = a(v^n)b(v^n), \qquad 0 \le n < N. \tag{13}$$

We see that multiplication $mod \ (x^N - 1)$ can be computed by N complex multiplications (13) along with some mechanism which translates between $\mathbb{C}\,[x]/(x^N - 1)$ and $\prod_{n=0}^{N-1} \mathbb{C}$. In fact, this mechanism is the N-point FFT, and (13) is a disguised form of the convolution theorem. To see this, observe that

$$a(1) = \sum_{n=0}^{N-1} a_n, \tag{14}$$

$$a(v) = \sum_{n=0}^{N-1} v^n a_n, \tag{15}$$

$$\vdots$$

$$a(v^{N-1}) = \sum_{n=0}^{N-1} v^{n(N-1)} a_n. \tag{16}$$

This implies

$$\begin{pmatrix} a(1) \\ a(v) \\ \vdots \\ a(v^{N-1}) \end{pmatrix} = F(N)\underline{a}, \tag{17}$$

where $\underline{a}$ is the vector of length N associated to the polynomial $a(x)$. Placing (17) into (13) we have

$$F(N)\underline{c} = (F(N)\underline{a})(F(N)\underline{b}), \tag{18}$$

where the right-hand side is a componentwise product. If $G(\underline{a})$ denotes the diagonal matrix defined in (6), we can rewrite (18) as

$$\underline{c} = F(N)^{-1}G(\underline{a})F(N)\underline{b}, \tag{19}$$

which is the convolution theorem.

6.3. Cook-Toom Algorithm

The derivation of the convolution theorem using the CRT admits important generalizations which can be used to design algorithms that for computing linear and cyclic convolution. The simplist is the Cook-Toom algorithm which we discuss in this section.

Take N distinct complex numbers

$$\{\alpha_0, \alpha_1, \cdots, \alpha_{N-1}\}, \tag{1}$$

and form a polynomial

$$m(x) = (x - \alpha_0)(x - \alpha_1)\cdots(x - \alpha_{N-1}). \tag{2}$$

We will begin by designing algorithms to compute polynomial multiplication *mod* $m(x)$ or equivalently multiplication in the quotient polynomial ring

$$\mathbb{C}\,[x]/m(x). \tag{3}$$

Applying the CRT as in the preceeding section, the mapping

$$a(x) \longrightarrow \begin{pmatrix} a(\alpha_0) \\ a(\alpha_1) \\ \cdots \\ a(\alpha_{N-1}) \end{pmatrix}, \tag{4}$$

establishes a ring-isomorphism

$$\mathbb{C}\,[x]/m(x) \cong \prod_{n=0}^{N-1} \mathbb{C}\,, \tag{5}$$

with the result that a polynomial $a(x)$ of degree $< N$ is uniquely determined by the values $a(\alpha_n)$, $0 \le n < N$. To see how to recover $a(x)$ from these values, write

$$a(\alpha_0) = a_0 + \alpha_0 a_1 + \cdots + \alpha_0^{N-1} a_{N-1},$$

$$\vdots$$

$$a(\alpha_{N-1}) = a_0 + \alpha_{N-1}a_1 + \cdots + \alpha_{N-1}^{N-1}a_{N-1}, \tag{6}$$

where $a(x) = \sum_{n=0}^{N-1} a_n x^n$. In matrix form, this becomes

$$\begin{bmatrix} a(\alpha_0) \\ a(\alpha_1) \\ \vdots \\ a(\alpha_{N-1}) \end{bmatrix} = W\underline{a}, \tag{7}$$

where $\underline{a}$ is the vector of componenets of $a(x)$ and W is the van de Monde matrix

$$W = \begin{bmatrix} 1 & \alpha_0 & \cdots & \alpha_0^{N-1} \\ 1 & \alpha_1 & \cdots & \alpha_1^{N-1} \\ \vdots & \vdots & & \vdots \\ 1 & \alpha_{N-1} & \cdots & \alpha_{N-1}^{N-1} \end{bmatrix}. \tag{8}$$

Since the elements α_n, $0 \leq n < N$, are distinct, the matrix W is invertible, so that we can recover $a(x)$ from

$$\underline{a} = W^{-1} \begin{bmatrix} a(\alpha_0) \\ a(\alpha_1) \\ \vdots \\ a(\alpha_{N-1}) \end{bmatrix}. \tag{9}$$

Consider two polynomials $a(x)$ and $b(x)$ in the quotient polynomial ring (3). Set

$$c(x) \equiv a(x)b(x) \, mod \, m(x). \tag{10}$$

Since $m(\alpha_n) = 0$, $0 \leq n < N$, we have

$$c(\alpha_n) = a(\alpha_n)b(\alpha_n), \qquad 0 \leq n < N. \tag{11}$$

Using (7) and (9) in (11), we have

$$\underline{c} = W^{-1}\big((W\underline{a})(W\underline{b})\big), \tag{12}$$

where $(W\underline{a})(W\underline{b})$ denotes componentwise multiplication. (12) generalizes the convolution theorem.

Example 1. Take $m(x) = x(x+1)$. Then

$$W = W^{-1} = \begin{bmatrix} 1 & 0 \\ 1 & -1 \end{bmatrix}.$$

Example 2. Take $m(x) = (x-1)(x+1)$. Then

$$W = F(2).$$

Example 3. Take $m(x) = x(x-1)(x+1)$. Then

$$W = \begin{bmatrix} 1 & 0 & 0 \\ 1 & 1 & 1 \\ 1 & -1 & 1 \end{bmatrix}.$$

Example 4. Take $m(x) = x(x-1)(x+1)(x-2)$. Then

$$W = \begin{bmatrix} 1 & 0 & 0 & 0 \\ 1 & 1 & 1 & 1 \\ 1 & -1 & 1 & -1 \\ 1 & 2 & 4 & 8 \end{bmatrix}.$$

(12) can be modified to design an algorithm for computing linear convolution. Consider polynomials $g(x)$ and $h(x)$ of degree $N-1$ and $M-1$, respectively. The linear convolution

$$s(x) = h(x)g(x), \tag{13}$$

has degree $L-1$, where $L = M + N - 1$. Denote by $\underline{g}$, $\underline{h}$ and $\underline{s}$ the vectors of sizes N, M and L, respectively, corresponding to the polynomials $g(x)$, $h(x)$ and $s(x)$. Take L distinct complex numbers α_l, $0 \le l < L$, and form the polynomial

$$m(x) = \prod_{l=0}^{L-1} (x - \alpha_l). \tag{14}$$

We call $m(x)$ a *reducing polynomial*. The design of an efficient algorithm depends, to a large extent, on the choice of a 'good' reducing polynomial. Define the submatrices of W in (8) by

$$W_M = \begin{bmatrix} 1 & \alpha_0 & \cdots & \alpha_0^{M-1} \\ 1 & \alpha_1 & \cdots & \alpha_1^{M-1} \\ \vdots & \vdots & & \vdots \\ 1 & \alpha_{L-1} & \cdots & \alpha_{L-1}^{M-1} \end{bmatrix},$$

and

$$W_N = \begin{bmatrix} 1 & \alpha_0 & \cdots & \alpha_0^{N-1} \\ 1 & \alpha_1 & \cdots & \alpha_1^{N-1} \\ \vdots & \vdots & & \vdots \\ 1 & \alpha_{L-1} & \cdots & \alpha_{L-1}^{N-1} \end{bmatrix}. \tag{15}$$

Since

$$deg(s(x)) = L - 1 < deg(m(x)) = L, \tag{16}$$

we have

$$s(x) = s(x) \bmod m(x), \tag{17}$$

which means that we can compute the linear convolution $s(x)$ by computing the product $h(x)g(x)$ in $\mathbb{C}[x]/m(x)$. In fact, (12) can be applied. Since

$$W_N \underline{g} = W\underline{g}, \qquad W_M \underline{h} = W\underline{h}, \tag{18}$$

we have

$$\underline{s} = W^{-1}\big((W_M \underline{h})(W_N \underline{g})\big). \tag{19}$$

In the examples that follow, we assume that

$$\alpha_l \in Z, \qquad 0 \le l < L, \tag{20}$$

which implies that W and W^{-1} are rational matrices. Computing the actions of W and W^{-1} require multiplications only by rational numbers. Real number multiplications occur in forming the component-wise product

$$(W_M \underline{h})(W_N \underline{g}) \tag{21}$$

which we can rewrite as diagonal matrix multiplication

$$D(\underline{h})W_N\underline{g}, \tag{22}$$

where

$$D(\underline{h}) = diag.(W_M\underline{h}). \tag{23}$$

In practice, the vector $\underline{h}$ is fixed over many computations and the diagonal matrix $D(\underline{h})$ is precomputed and is not counted in the arithmetic cost. In this case, computing the linear convolution can be carried out in the followiung three operations.

1. Compute $W_N\underline{g}$. This part requires $(N-1)L$ additions and LN multiplications by rational numbers.

2. Compute $D(\underline{h})W_N\underline{g}$. This part requires L multiplications.

3. Compute $W^{-1}(D(\underline{h})W_N\underline{g})$. This part reqquires $L(L-1)$ additions and L^2 multiplications by rational numbers.

<u>Summary</u>: computing linear convolution with $\underline{h}$ fixed and $W_M(\underline{h})$ precomputed requires

$$(L+N-2)L \qquad \underline{\text{additions}},$$

$$(L+N)L \quad \underline{\text{multiplications by rationals}},. \tag{24}$$

$$L \qquad \underline{\text{multiplications}}$$

This should be compared with the straightforward computation of $M \times N$ linear convolution which requires

$$(N-1)(M-1) \qquad \underline{\text{additions}},$$

$$NM \qquad \underline{\text{multiplications}} \tag{25}$$

The arithmetic described in (24) is for the general case. Significant reduction occurs if the numbers α_l, $0 \leq l < L$, are carefully chosen.

Example 5. Consider 2×2 linear convolution, and take $m(x) = x(x-1)(x+1)$, then

$$
\underline{s} = \begin{bmatrix} 1 & 0 & 0 \\ 0 & \frac{1}{2} & -\frac{1}{2} \\ -1 & \frac{1}{2} & \frac{1}{2} \end{bmatrix} \left(\left(\begin{bmatrix} 1 & 0 \\ 1 & 1 \\ 1 & -1 \end{bmatrix} \underline{h} \right) \left(\begin{bmatrix} 1 & 0 \\ 1 & 1 \\ 1 & -1 \end{bmatrix} \underline{g} \right) \right).
$$

We can write this out in the following sequence of steps:

0. Precompute

$$
H_0 = h_0, \qquad H_1 = h_0 + h_1, \qquad H_2 = h_0 - h_1.
$$

1. Compute

$$
G_0 = g_0, \qquad G_1 = g_0 + g_1, \qquad G_2 = g_0 - g_1.
$$

2. Compute

$$
S_0 = H_0 G_0, \qquad S_1 = H_1 G_1, \qquad S_2 = H_2 G_2.
$$

3. Compute

$$
s_0 = S_0, \qquad s_1 = \frac{1}{2}(S_1 - S_2), \qquad s_2 = -S_0 + \frac{1}{2}(S_1 + S_2).
$$

If we compute $\frac{1}{2}H_1$ and $\frac{1}{2}H_2$, in the precomputation stage, then multiplication by $\frac{1}{2}$ in step 3 can be eliminated. We see that 5 additions and 3 multiplications are required to carry out the computation compared to 1 addition and 4 multiplications by straightforward methods. As is typical, multiplications are reduced at the expense of additions.

A better algorithm can be produced using the following modification. Consider, again, the linear convolution, $s(x) = h(x)g(x)$, of polynomials $h(x)$ and $g(x)$ of degrees $M-1$ and $N-1$, respectively. Take $L-1$ distinct numbers

$$\alpha_l, \qquad 0 \le l < L-1, \tag{26}$$

and form the polynomial

$$m'(x) = (x - \alpha_0)(x - \alpha_1) \cdots (x - \alpha_{L-2}). \tag{27}$$

Compute

$$c(x) \equiv h(x)g(x) \qquad mod\ m'(x). \tag{28}$$

Since

$$deg(s(x)) = deg(m'(x)), \tag{29}$$

$s(x) \ne c(x)$, but we can recover $s(x)$ from $c(x)$ by the formula

$$s(x) = c(x) + h_{M-1}g_{N-1}m'(x). \tag{30}$$

Now we compute $s(x)$ in two stages:

1. Compute $c(x) \equiv h(x)g(x)\ mod\ m'(x)$,
2. Compute $s(x)$ by (30).

The modification above reduces the required additions without any change in the required multiplications. Computing $s(x)$ by the above two stages is denoted by

$$s(x) \equiv h(x)g(x) \qquad mod\ m'(x)(x - \infty). \tag{31}$$

Example 6. Consider, again, 2×2 linear convolution by taking $m(x) = x(x+1)(x-\infty)$ and computing

$$s(x) \equiv h(x)g(x)\ mod\ m(x).$$

First compute

$$c(x) \equiv h(x)g(x)\ mod\ x(x+1),$$

by

$$\underline{c} = \begin{bmatrix} 1 & 0 \\ 1 & -1 \end{bmatrix} \left(\left(\begin{bmatrix} 1 & 0 \\ 1 & -1 \end{bmatrix} \underline{h} \right) \left(\begin{bmatrix} 1 & 0 \\ 1 & -1 \end{bmatrix} \underline{g} \right) \right).$$

Writing this out, we have the following sequence of operations:

$$0. \quad H_0 = h_0, \quad H_1 = h_0 - h_1$$

$$1. \quad G_0 = g_0, \quad G_1 = g_0 - g_1$$

$$2. \quad S_0 = H_0 G_0, \quad S_1 = H_1 G_1$$

$$3. \quad c_0 = S_0, \quad c_1 = S_0 - S_1$$

We assume step 0 is pre-computed. Steps 1–3 require 2 additions and 2 multiplications. We complete the computation by the following sequence of operations.

$$4. \quad s_2 = h_1 g_1$$

$$5. \quad s_0 = c_0, \quad s_1 = c_1 + s_2.$$

In the above algorithm, 3 additions and 3 multiplications are required reducing the arithmetic cost by 2 additions as compared to example 5.

Example 7. Consider 2×3 linear convolution and let

$$m(x) = x(x - 1)(x + 1)(x - \infty).$$

The computation of linear convolution

$$s(x) = h(x)g(x),$$

where $deg(h(x)) = 1$ and $deg(g(x)) = 2$, can be carried out by first computing

$$c(x) \equiv h(x)g(x) \bmod x(x - 1)(x + 1),$$

and then using the formula

$$s(x) = c(x) + h_1 g_2 x(x - 1)(x + 1).$$

The first part is given by

$$
\underline{c} =
\begin{bmatrix} 1 & 0 & 0 \\ 0 & 1 & -1 \\ -1 & 1 & 1 \end{bmatrix}
\begin{bmatrix} 1 & 0 & 0 \\ 0 & \frac{1}{2} & 0 \\ 0 & 0 & \frac{1}{2} \end{bmatrix}
\left(\begin{bmatrix} 1 & 0 \\ 1 & 1 \\ 1 & -1 \end{bmatrix} \underline{h} \right)
$$

$$
\left(\begin{bmatrix} 1 & 0 & 0 \\ 1 & 1 & 1 \\ 1 & -1 & 1 \end{bmatrix} \underline{g} \right).
$$

We carry out this computation as follows:

0. $H_0 = h_0, \; H_1 = \dfrac{h_0 + h_1}{2}, \; H_2 = \dfrac{h_0 - h_1}{2} \; (pre-compute).$

1. $G_0 = g_0, \quad G_1 = g_0 + g_1 + g_2, \quad G_2 = g_0 - g_1 + g_2.$

2. $S_0 = H_0 G_0, \quad S_1 = H_1 G_1, \quad S_2 = H_2 G_2.$

3. $c_0 = S_0, \quad c_1 = S_1 - S_2, \quad c_2 = -S_0 + S_1 + S_2.$

This part requires 6 additions and 3 multiplications(as before, multiplications by $\frac{1}{2}$ have been placed in precomputation stage).

We complete the computation by

4. $s_3 = h_1 g_2.$

5. $s_0 = c_0, \quad s_1 = c_1 - s_3, \quad s_2 = c_2.$

Both steps can be given in one matrix equation as follows

$$
\underline{s} =
\begin{bmatrix} 1 & 0 & 0 & 0 \\ 0 & 1 & -1 & -1 \\ -1 & 1 & 1 & 0 \\ 0 & 0 & 0 & 1 \end{bmatrix}
\begin{bmatrix} 1 & 0 & 0 & 0 \\ 0 & \frac{1}{2} & 0 & 0 \\ 0 & 0 & \frac{1}{2} & 0 \\ 0 & 0 & 0 & 1 \end{bmatrix}
\left(\begin{bmatrix} 1 & 0 \\ 1 & 1 \\ 1 & -1 \\ 0 & 1 \end{bmatrix} \underline{h} \right)
$$

$$
\left(\begin{bmatrix} 1 & 0 & 0 \\ 1 & 1 & 1 \\ 1 & -1 & 1 \\ 0 & 0 & 1 \end{bmatrix} \underline{g} \right).
$$

The small size linear convolution described in the above examples can be efficiently computed by the Cook-Toom algorithm. The factors of the reducing polynomials have roots 0, ± 1 with the result that the matrcies W_M and W_N have coefficients 0, ± 1. The matrix W^{-1} is more complicated but the rational multiplications can be carried out in the pre-computation stage. This is a general result which will be discussed in section 5. As the size of the problem grows, the roots of the reducing polynomials must contain large integers which appear along with their powers in the matrices W_M and W_N. If the large integer multiplications are carried out by additions, then as the size of the problem grows, the number of required additions grows too large for practical implementation. In any case, the computation becomes less stable as the size grows [3]. In the next section, we will present efficient larger size algorithms using generalization of Cook-Toom algorithm.

Linear convolution can be used to compute multiplication in quotient polynomial rings. In section 4, we will extensively use this approach to cyclic convolution algorithms [5].

Example 8. We want to compute

$$c_l(x) \equiv h(x)g(x) \bmod m_l(x),$$

where

$$m_1(x) = x^2,$$
$$m_2(x) = x^2 + 1,$$
$$m_3(x) = x^2 - 1,$$
$$m_4(x) = x^2 + x + 1,$$
$$m_5(x) = x^2 - x + 1.$$

Computing first linear convolution $s(x) = h(x)g(x)$ by the algorithm designed in example 6, we have

$$\underline{s} = \begin{bmatrix} 1 & 0 & 0 \\ 1 & -1 & 1 \\ 0 & 0 & 1 \end{bmatrix} \left(\left(\begin{bmatrix} 1 & 0 \\ 1 & -1 \\ 0 & 1 \end{bmatrix} \underline{h} \right) \left(\begin{bmatrix} 1 & 0 \\ 1 & -1 \\ 0 & 1 \end{bmatrix} \underline{g} \right) \right).$$

The operation, $mod\ m_l(x)$, can be viewed as matrix multiplication. Set

$$A = \begin{bmatrix} 1 & 0 \\ 1 & -1 \\ 0 & 1 \end{bmatrix},$$

we have

$$\underline{c}_1 = \begin{bmatrix} 1 & 0 & 0 \\ 1 & -1 & 1 \end{bmatrix} ((A\underline{h})\,(A\underline{f}))\,.$$

Continuing in this way, we have

$$\underline{c}_2 = \begin{bmatrix} 1 & 0 & -1 \\ 1 & -1 & 1 \end{bmatrix} ((A\underline{h})\,(A\underline{g}))\,,$$

$$\underline{c}_3 = \begin{bmatrix} 1 & 0 & 1 \\ 1 & -1 & 1 \end{bmatrix} ((A\underline{h})\,(A\underline{g}))\,,$$

$$\underline{c}_4 = \begin{bmatrix} 1 & 0 & -1 \\ 1 & -1 & 0 \end{bmatrix} ((A\underline{h})\,(A\underline{g}))\,,$$

$$\underline{c}_5 = \begin{bmatrix} 1 & 0 & -1 \\ 1 & -1 & 2 \end{bmatrix} ((A\underline{h})\,(A\underline{g}))\,.$$

6.4. Winograd Small Convolution Algorithm

The Cook-Toom algorithm uses the CRT relative to a reducing polynomial $m(x)$ constructed from linear factors having integer coefficients. Although non-rational multiplications are kept to a minimum, additions grow rapidly as the size of the computation increases. A major part of these additions are needed to carry out the rational multiplications coming from the integer coefficients of the linear factors. Of major importance is the numerical stability of the computation [3].

By applying the CRT more generally, Winograd designed algorithms which could efficiently handle a larger collection of small size convolutions. The growth in the number of required additions is not as rapid as in the Cook-Toom algorithm while the cost in the number of required multiplications increases modestly.

Consider a reducing polynomial

$$m(x) = m_1(x)m_2(x)\cdots m_r(x), \tag{1}$$

where $m_l(x)$, $0 \leq l \leq r$, are relatively prime. We do not require that these polynomials be linear. This leads to the possibility of building reducing polynomials $m(x)$ of higher degrees than before and still have factors with small integer coefficients. As we saw in the preceeding section the coefficients of the factors of the reducing polynomials become multipliers in the corresponding algorithm. If these coefficients are small integers, then these multiplications can be carried out by a small number of additions. As the size of these integers grows the number of required additions grows.

Suppose $m(x) = m_1(x)m_2(x)$ where $m_1(x)$ and $m_2(x)$ are relatively prime. The extension to more relatively prime factors follows easily. We want to compute multiplication in the polynomial ring

$$\mathbb{C}\,[x]/m(x). \tag{2}$$

By the CRT, we can carry out this computation as follows. Suppose

$$deg(m(x)) = N, \quad deg(m_l(x)) = N_l, \quad l = 1,2.$$

Take polynomials $h(x)$ and $g(x)$ of degree $< N$. We want to compute

$$c(x) \equiv h(x)g(x) \bmod m(x). \tag{3}$$

1. Compute the reduced polynomials

$$h^{(k)}(x) \equiv h(x) \bmod m_k(x), \quad k = 1,2.$$

$$h^{(k)}(x) \equiv g(x) \bmod m_k(x), \quad k = 1, 2.$$

2. Compute the products

$$c^{(l)}(x) \equiv h^{(l)}(x)g^{(k)}(x) \bmod m_k(x), \quad k = 1, 2.$$

The CRT guarantees that $c(x)$ is uniquely determined by the polynomials $c^{(k)}(x)$, $k = 1, 2$, and prescribes a method for its computation. The unique system of idempotents

$$e_1(x), \qquad e_2(x), \tag{4}$$

corresponding to the factorization $m(x) = m_1(x)m_2(x)$ satisfies

$$e_k(x) \equiv 1 \bmod m_k(x), \quad k = 1, 2, \tag{5}$$

$$e_l(x) \equiv 0 \bmod m_k(x), \quad l, k = 1, 2, l \neq k. \tag{6}$$

Then

$$1 \equiv e_1(x) + e_2(x) \bmod m(x),$$

and

$$c(x) \equiv c^{(1)}(x)e_1(x) + c^{(2)}(x)e_2(x). \tag{7}$$

To complete the computation of $c(x)$ from (7), we require the following steps :

3. Compute the products

$$c_k(x) \equiv c^{(k)}(x)E_k(x) \bmod m_k(x), \ k = 1, 2.$$

4. Compute the sum

$$c(x) = c_1(x) + c_2(x).$$

In the second stage of the algorithm, we compute multiplications in the polynomial rings

$$\mathbf{C}\,[x]/m_l(x), \quad l = 1, 2. \tag{8}$$

In part, multiplications in the polynomial ring (2) have been replaced by multiplications in the polynomial rings (8). Efficient small size algorithms computing multiplications in (8) provide building blocks for computing multiplications in (2).

In the previous section, we designed algorithms for computing linear convolution and multiplication in quotient polynomial rings of the form

$$\underline{c} = C \left((A\underline{h}) (B\underline{g}) \right), \tag{9}$$

where $\underline{h}$, $\underline{g}$ are input vectors, $\underline{c}$ is output vector and A, B and C are matrices corresponding to W_M, W_N and W^{-1}. Such algorithms are *bilinear algorithms*. We will now see how to piece together bilinear algorithms computing multiplication in the polynomial ring $\mathbb{C}[x]/m(x)$. This permits efficient small algorithms to be used as building blocks in larger size algorithms.

Suppose bilinear algorithms compute the products,

$$c^{(k)}(x) \equiv h^{(k)}(x)g^{(k)}(x) \ mod \ m_k(x), \quad k = 1, 2,$$

$$\underline{c}^{(k)} = C_k \left(\left(A_k \underline{h}^{(k)} \right) \left(B_k \underline{g}^{(k)} \right) \right). \tag{10}$$

We assume that A_k and B_k are $N \times N_k$ matrices and C_k is an $N_k \times N$ matrix. The operation $mod \ m_k(x)$ can be computed by matrix multiplication

$$\underline{h}^{(k)} = M_k \underline{h}, \tag{11}$$

$$\underline{g}^{(k)} = M_k \underline{g}, \tag{12}$$

where M_k is an $N_k \times N$ matrix having coefficients determined by $m_j(x)$. Set

$$A = \begin{bmatrix} A_1 \\ A_2 \end{bmatrix}, \qquad B = \begin{bmatrix} B_1 \\ B_2 \end{bmatrix}, \qquad M = \begin{bmatrix} M_1 \\ M_2 \end{bmatrix}, \tag{13}$$

and set

$$AM = \begin{bmatrix} A_1 M_1 \\ A_2 M_2 \end{bmatrix}, \qquad BM = \begin{bmatrix} B_1 M_1 \\ B_2 M_2 \end{bmatrix}, \tag{14}$$

$$C = \begin{bmatrix} C_1 & 0 \\ 0 & C_2 \end{bmatrix}, \tag{15}$$

then

$$\begin{bmatrix} \underline{c}^{(1)} \\ \underline{c}^{(2)} \end{bmatrix} = C(AM\underline{h})(BM\underline{g}). \tag{16}$$

The vectors $\underline{c}^{(1)}$ and $\underline{c}^{(2)}$ determine the polynomials $c^{(1)}(x)$ and $c^{(2)}(x)$ which must now be put together using the idempotents. Multiplication by $e_k(x) \bmod m(x)$ can be described by a $N \times N_k$ matrix E_k, $k = 1, 2$. We have

$$\underline{c}_k = E_k \underline{c}^{(k)}, \qquad k = 1, 2, \tag{17}$$

then

$$\underline{c} = EC((AM\underline{h})(BM\underline{g})), \tag{18}$$

where

$$E = E_1 E_2.$$

The efficiency of (18) depends on two factors: the efficiency of the small bilinear algorithms (10) and the efficiency of how these building blocks are put together. We assume throughout that the factors $m_1(x)$ and $m_2(x)$ contain only small integer coefficients. Then M has only small integer coefficients. Although the matrix E has rational coefficients, as we will see in section 6, its action can frequently be computed in a precomputation stage.

Example 1. Take

$$m(x) = x(x^2 + 1).$$

We will use (18) to compute

$$c(x) \equiv h(x)f(x) \bmod m(x),$$

where the bilinear algorithm computing multiplication $mod\ (x^2 + 1)$ is taken from example 8 of the preceeding section. First with

$$m_1(x) = x, \qquad m_2(x) = x^2 + 1,$$

we have

$$M_1 = [1 \ \ 0 \ \ 0], \qquad M_2 = \begin{bmatrix} 1 & 0 & -1 \\ 0 & 1 & 0 \end{bmatrix}.$$

From example (3.8),

$$A_2 = B_2 = \begin{bmatrix} 1 & 0 \\ 1 & -1 \\ 0 & 1 \end{bmatrix},$$

$$C_2 = \begin{bmatrix} 1 & 0 & -1 \\ 1 & -1 & 1 \end{bmatrix}.$$

We can see directly that,

$$A_1 = C_1 = B_1 = [\, 1 \,].$$

Then

$$AM = BM = \begin{bmatrix} 1 & 0 & 0 \\ 1 & 0 & -1 \\ 1 & -1 & -1 \\ 0 & 1 & 0 \end{bmatrix},$$

$$C = \begin{bmatrix} 1 & 0 & 0 & 0 \\ 0 & 1 & 0 & -1 \\ 0 & 1 & -1 & 0 \end{bmatrix}.$$

The idempotents are given by

$$e_1(x) = x^2 + 1, \qquad e_2(x) = -x^2,$$

from which it follows that

$$E_1 = \begin{bmatrix} 1 \\ 0 \\ 1 \end{bmatrix}, \qquad E_2 = \begin{bmatrix} 0 & 0 \\ 0 & 1 \\ -1 & 0 \end{bmatrix}.$$

Direct computation shows that

$$EC = \begin{bmatrix} 1 & 0 & 0 & 0 \\ 0 & 1 & -1 & 1 \\ 1 & -1 & 0 & 1 \end{bmatrix}.$$

Thus

$$\underline{c} = \begin{bmatrix} 1 & 0 & 0 & 0 \\ 0 & 1 & -1 & 1 \\ 1 & -1 & 0 & 1 \end{bmatrix} \left(\begin{bmatrix} 1 & 0 & 0 \\ 1 & 0 & -1 \\ 1 & -1 & -1 \\ 0 & 1 & 0 \end{bmatrix} \underline{h} \right) \left(\begin{bmatrix} 1 & 0 & 0 \\ 1 & 0 & -1 \\ 1 & -1 & -1 \\ 0 & 1 & 0 \end{bmatrix} \underline{g} \right).$$

Example 2. Take

$$m(x) = (x+1)(x^2 + x + 1),$$

where

$$m_1(x) = x + 1, \qquad m_2(x) = x^2 + x + 1.$$

Then

$$M_1 = [1 \quad -1 \quad 1], \qquad M_2 = \begin{bmatrix} 1 & 0 & -1 \\ 0 & 1 & -1 \end{bmatrix}.$$

Directly

$$A_1 = B_1 = C_1 = [\,1\,].$$

By example (3.8), we can take

$$A_2 = B_2 = \begin{bmatrix} 1 & 0 \\ 1 & -1 \\ 0 & 1 \end{bmatrix},$$

$$C_2 = \begin{bmatrix} 1 & 0 & -1 \\ 1 & -1 & 0 \end{bmatrix}.$$

The idempotents are given by

$$e_1(x) = x^2 + x + 1, \qquad e_2(x) = -(x^2 + x),$$

implying that

$$E_1 = \begin{bmatrix} 1 \\ 1 \\ 1 \end{bmatrix}, \qquad E_2 = \begin{bmatrix} 0 & 1 \\ -1 & 2 \\ -1 & 1 \end{bmatrix}.$$

Putting this all together in (18), we have

$$\underline{c} = \begin{bmatrix} 1 & 1 & -1 & 0 \\ 1 & 1 & -2 & 1 \\ 1 & -1 & 0 & 1 \end{bmatrix} \left(\begin{bmatrix} 1 & -1 & 1 \\ 1 & 0 & -1 \\ 1 & -1 & 0 \\ 0 & 1 & -1 \end{bmatrix} \underline{h} \right) \left(\begin{bmatrix} 1 & -1 & 1 \\ 1 & 0 & -1 \\ 1 & -1 & 0 \\ 0 & 1 & -1 \end{bmatrix} \underline{g} \right).$$

<u>Example</u> 3. We design an algorithm computing multiplication mod $m(x)$ where

$$m(x) = m_1(x)m_2(x),$$

$$m_1(x) = x(x^2 + 1), \qquad m_2(x) = (x+1)(x^2 + x + 1).$$

From the preceeding two examples, we can compute multiplication mod $m_l(x)$, $l = 1, 2$, by taking

$$A_1 = B_1 = \begin{bmatrix} 1 & 0 & 0 \\ 1 & 0 & -1 \\ 1 & -1 & -1 \\ 0 & 1 & 0 \end{bmatrix},$$

$$C_1 = \begin{bmatrix} 1 & 0 & 0 & 0 \\ 0 & 1 & -1 & 1 \\ 1 & -1 & 0 & 1 \end{bmatrix},$$

$$A_2 = B_2 = \begin{bmatrix} 1 & -1 & 1 \\ 1 & 0 & -1 \\ 1 & -1 & 0 \\ 0 & 1 & -1 \end{bmatrix},$$

$$C_2 = \begin{bmatrix} 1 & 1 & -1 & 0 \\ 1 & 1 & -2 & 1 \\ 1 & -1 & 0 & 1 \end{bmatrix}.$$

Directly

$$M_1 = \begin{bmatrix} 1 & 0 & 0 & 0 & 0 & 0 \\ 0 & 1 & 0 & -1 & 0 & 1 \\ 0 & 0 & 1 & 0 & -1 & 0 \end{bmatrix},$$

$$M_2 = \begin{bmatrix} 1 & 0 & 0 & -1 & 2 & -2 \\ 0 & 1 & 0 & -2 & 3 & -2 \\ 0 & 0 & 1 & -2 & 2 & -1 \end{bmatrix},$$

then,

$$AM = BM = \begin{bmatrix} 1 & 0 & 0 & 0 & 0 & 0 \\ 1 & 0 & -1 & 0 & 1 & 0 \\ 1 & -1 & -1 & 1 & 1 & -1 \\ 0 & 1 & 0 & -1 & 0 & 1 \\ 1 & -1 & 1 & -1 & 1 & -1 \\ 1 & 0 & -1 & 1 & 0 & -1 \\ 1 & -1 & 0 & 1 & -1 & 0 \\ 0 & 1 & -1 & 0 & 1 & -1 \end{bmatrix}.$$

The idempotents are given by

$$e_1(x) = \frac{1}{2}(3x^5 + 5x^4 + 6x^3 + 5x^2 + 3x + 2),$$

$$e_2(x) = -\frac{1}{2}(3x^5 + 5x^4 + 6x^3 + 3x^2 + 3x),$$

and since

$$m(x) = x^6 + 2x^5 + 3x^4 + 3x^3 + 2x^2 + x,$$

we have that

$$E_1 = \frac{1}{2}\begin{bmatrix} 2 & 0 & 0 \\ 3 & -1 & 1 \\ 5 & -3 & 1 \\ 6 & -4 & 0 \\ 5 & -3 & -1 \\ 3 & -1 & -1 \end{bmatrix}, \qquad E_2 = -\frac{1}{2}\begin{bmatrix} 0 & 0 & 0 \\ 3 & -3 & 1 \\ 5 & -3 & -1 \\ 6 & -4 & 0 \\ 5 & -3 & -1 \\ 3 & -1 & -1 \end{bmatrix}.$$

Direct computation shows that $C' = EC$ is given by

$$C' = \frac{1}{2}\begin{bmatrix} 2 & 0 & 0 & 0 & 0 & 0 & 0 & 0 \\ 4 & -2 & 1 & 0 & -1 & 1 & -3 & 2 \\ 6 & -4 & 3 & -2 & -1 & -3 & -1 & 4 \\ 6 & -4 & 4 & -4 & -2 & -2 & -2 & 4 \\ 4 & -2 & 3 & -4 & -1 & -3 & -1 & 4 \\ 2 & 0 & 1 & -2 & -1 & -3 & 1 & 1 \end{bmatrix}.$$

Then by (18),

$$\underline{c} = C'((AM\underline{h})(AM\underline{g})).$$

6.5. Linear and Cyclic Convolutions

The methods of section 4 decompose the computation of polynomial multiplication $mod\ m(x)$ into small size computations of polynomial multiplication $mod\ m_1(x)$ and $m_2(x)$, where $m(x) = m_1(x)m_2(x)$, $m_1(x)$ and $m_2(x)$ are relatively prime. We can apply these ideas to linear and cyclic convolution to decompose a large size problem into several small size problems. As we will see, algorithms computing linear convolution, cyclic convolution and multiplication modules a polynomial can be used as a part of other algorithms of the same type. This permits large size problems to be successively decomposed into smaller and smaller problems.

Example 1. Consider 2×2 linear convolution. Using the algorithm given by (4.18) for the reducing polynomial $m(x) = x(x^2 - 1)$ leads to the same algorithm as that designed in section 3.

Example 2. Consider 2×3 linear convolution

$$s(x) = g(x)h(x),$$

where

$$g(x) = g_0 + g_1 x, \qquad h(x) = h_0 + h_1 x + h_2 x^2.$$

First compute the product

$$c(x) \equiv g(x)h(x)\ mod\ x(x^2 + 1)$$

by the algorithm of example (4.1). Then we compute $s(x)$ by

$$s(x) = c(x) + g_1 h_2 x(x^2 + 1).$$

In the next few examples, two 3×3 linear convolutions will be derived. The first is based on the Cook-Toom algorithm while the second follows from the methods of section 4.

Example 3. Consider 3×3 linear convolution

$$s(x) = g(x)h(x),$$

for polynomial $g(x)$ and $h(x)$ of size 3. Compute

$$c(x) \equiv g(x)h(x) \bmod x(x-1)(x+1)(x-2)$$

by the Cook-Toom algorithm. Then

$$s(x) = c(x) + g_2 h_2 x(x-1)(x+1)(x-2).$$

Working this out, we have

$$\underline{s} = \frac{1}{4}C((A\underline{g})(A\underline{h})),$$

where

$$A = \begin{bmatrix} 1 & 0 & 0 \\ 1 & 1 & 1 \\ 1 & -1 & 1 \\ 1 & 2 & 4 \\ 0 & 0 & 1 \end{bmatrix},$$

$$C = \begin{bmatrix} 4 & 0 & 0 & 0 & 0 \\ 0 & 2 & -2 & 0 & 8 \\ -7 & 5 & 3 & -1 & -4 \\ 3 & -3 & -1 & 1 & -8 \\ 0 & 0 & 0 & 0 & 4 \end{bmatrix}.$$

Before designing the second 3×3 linear convolution algorithm, we will design a 4-point cyclic convolution algorithm which will then be used to design a 3×3 linear convolution algorithm having slightly more multiplications but significantly fewer additions. We also note that the convolution theorem can be used to efficiently compute 4-point cyclic convolution.

Example 4. Consider 4-point cyclic convolution

$$c(x) \equiv g(x)h(x) \ mod \ (x^4 - 1).$$

Using the factorization

$$x^4 - 1 = (x^2 + 1)(x^2 - 1),$$

in (18), we have

$$M_1 = \begin{bmatrix} 1 & 0 & -1 & 0 \\ 0 & 1 & 0 & -1 \end{bmatrix}, \quad M_2 = \begin{bmatrix} 1 & 0 & 1 & 0 \\ 0 & 1 & 0 & 1 \end{bmatrix}.$$

Compute multiplication $mod \ (x^2 + 1)$ and multiplication $mod \ (x^2 - 1)$ by example (3.8). Then

$$A_1 = A_2 = B_1 = B_2 = \begin{bmatrix} 1 & 0 \\ 1 & -1 \\ 0 & 1 \end{bmatrix},$$

$$C_1 = \begin{bmatrix} 1 & 0 & -1 \\ 1 & -1 & 1 \end{bmatrix}, \quad C_2 = \begin{bmatrix} 1 & 0 & 1 \\ 1 & -1 & 1 \end{bmatrix}.$$

The idempotents are given by

$$E_1(x) = -\frac{1}{2}(x^2 - 1), \qquad E_2(x) = \frac{1}{2}(x^2 + 1),$$

and we have

$$e_1 = \frac{1}{2} \begin{bmatrix} 1 & 0 \\ 0 & 1 \\ -1 & 0 \\ 0 & -1 \end{bmatrix}, \qquad e_2 = \frac{1}{2} \begin{bmatrix} 1 & 0 \\ 0 & 1 \\ 1 & 0 \\ 0 & 1 \end{bmatrix}.$$

Putting all these together,

$$\underline{c} = \frac{1}{2} C((A\underline{g})(A\underline{h})),$$

where

$$C = \begin{bmatrix} 1 & 0 & -1 & 1 & 0 & 1 \\ 1 & -1 & 1 & 1 & -1 & 1 \\ -1 & 0 & 1 & 1 & 0 & 1 \\ -1 & 1 & -1 & 1 & -1 & 1 \end{bmatrix},$$

$$A = \begin{bmatrix} 1 & 0 & -1 & 0 \\ 1 & -1 & -1 & 1 \\ 0 & 1 & 0 & -1 \\ 1 & 0 & 1 & 0 \\ 1 & -1 & 1 & -1 \\ 0 & 1 & 0 & 1 \end{bmatrix}.$$

Example 5. Consider 3×3 linear convolution

$$s(x) = g(x)h(x).$$

First compute the 4-point cyclic convolution

$$c(x) \equiv g(x)h(x) \bmod (x^4 - 1),$$

by the algorithm designed in section 4. We note that since the degree of $g(x)$ and $h(x)$ is equal to two, we can rewrite example 4 as

$$\underline{c} = \frac{1}{2}C'((A'\underline{g})(A'\underline{h})),$$

where

$$C' = \begin{bmatrix} 1 & 0 & -1 & 0 & 1 \\ 1 & -1 & 1 & -1 & 1 \\ -1 & 0 & 1 & 0 & 1 \\ -1 & 1 & -1 & -1 & 1 \end{bmatrix},$$

$$A' = \begin{bmatrix} 1 & 0 & -1 \\ 1 & -1 & 0 \\ 0 & 1 & 0 \\ 1 & 0 & 1 \\ 1 & -1 & 1 \\ 0 & 1 & 0 \end{bmatrix}.$$

We can now compute $s(x)$ by

$$s(x) = c(x) + g_2 h_2(x^4 - 1).$$

As compared to example 3, we see that now all coefficients are $0, \pm 1$ while in example 3, 'large' integers appear in the matrices.

Example 6. Consider 4×4 linear convolution

$$s(x) = g(x)h(x).$$

First compute

$$c(x) \equiv g(x)h(x) \ mod \ m(x),$$

where the reducing polynomial is

$$m(x) = x(x+1)(x^2+1)(x^2+x+1).$$

Then

$$s(x) = c(x) + g_3 h_3 m(x).$$

To compute $c(x)$, we use the algorithm designed in example 3 of section 4. Since

$$deg(g(x)) = deg(h(x)) = 3,$$

we have

$$\underline{c} = C'((A'\underline{g})(A'\underline{h})),$$

where

$$A' = \begin{bmatrix} 1 & 0 & 0 & 0 \\ 1 & 0 & -1 & 0 \\ 1 & -1 & -1 & 0 \\ 0 & 1 & 0 & -1 \\ 1 & -1 & 1 & -1 \\ 1 & 0 & -1 & 1 \\ 1 & -1 & 0 & 1 \\ 0 & 1 & -1 & 0 \end{bmatrix},$$

and C' is as given in example 3 of section 4.

Consider N-point cyclic convolution,

$$c(x) \equiv g(x)h(x) \ mod \ (x^N - 1). \tag{1}$$

If an efficient N–point FT is available, then the convolution theorem is usually the best approach for computing N–point cyclic convolution. The algorithms of section 4 can also be called upon. For instance, take the factorization,

$$x^N - 1 = \prod_{k=0}^{K-1} \phi_k(x), \tag{2}$$

where the polynomials $\phi_k(x)$, $0 \le k < K$, are the prime factors of $x^N - 1$ over the rational field $\mathbf{Q}$. These polynomials are usually called *cyclotomic polynomials*. If,

$$g^{(k)} \equiv g(x) \bmod \phi_k(x), \tag{3}$$

$$h^{(k)} \equiv h(x) \bmod \phi_k(x), \qquad 0 \le k < K, \tag{4}$$

then the cyclic convolution $c(x)$ can be found from the products,

$$c^{(k)}(x) \equiv g^{(k)}(x)h^{(k)}(x) \bmod \phi_k(x) \qquad 0 \le k < K, \tag{5}$$

by the formula,

$$c(x) \equiv \sum_{k=0}^{K-1} c^{(k)}(x)E_k(x), \tag{6}$$

where,

$$\{E_k(x) : \ 0 \le k < K\}$$

is the unique system of idempotents corresponding to the factorization (2).

As discussed in section 4, choosing a factorization over the rational field $\mathbf{Q}$ implies that the only multiplications required to carry out the algorithm are those given in (5). We continue to assume that the factorization is over $\mathbf{Q}$, but point out that factorization over other field can lead to efficient algorithms. This will be the case if multiplication by elements from the field can be efficiently implemented. For example, the field $Q(i)$ of Gaussian numbers consisting of all complex numbers $a + ib$, a and b rational, is frequently taken. The value of 'extending' the field of the factorization is that the prime factors are of smaller degree.

In the following examples, we will work out algorithms following the above approach. The multiplications (5) will be computed by first passing through linear convolution.

<u>**Example 7**</u>. Consider 3-point cyclic convolution,

$$c(x) \equiv g(x)h(x) \bmod (x^3 - 1).$$

The factorization (2) is given by

$$x^3 - 1 = (x - 1)(x^2 + x + 1).$$

Example (3.8) provides the bilinear algorithm for computing multiplication modulo $x^2 + x + 1$. We have

$$M_1 = [1 \quad 1 \quad 1], \qquad M_2 = \begin{bmatrix} 1 & 0 & -1 \\ 0 & 1 & -1 \end{bmatrix},$$

$$A_1 = B_1 = C_1 = [\, 1 \,],$$

$$A_2 = B_2 = \begin{bmatrix} 1 & 0 \\ 1 & -1 \\ 0 & 1 \end{bmatrix},$$

$$C_2 = \begin{bmatrix} 1 & 0 & -1 \\ 1 & -1 & 0 \end{bmatrix}.$$

The idempotents are given by

$$e_1(x) = \frac{1}{3}(x^2 + x + 1), \qquad e_2(x) = -\frac{1}{3}(x^2 + x - 2),$$

and

$$E_1 = \frac{1}{3}\begin{bmatrix} 1 \\ 1 \\ 1 \end{bmatrix}, \qquad E_2 = \begin{bmatrix} 2 & -1 \\ -1 & 2 \\ -1 & -1 \end{bmatrix}.$$

By (4.18), we have the bilinear algorithm

$$\underline{c} = \frac{1}{3}C'((A'\underline{g})(A'\underline{h})),$$

where

$$A' = \begin{bmatrix} 1 & 1 & 1 \\ 1 & 0 & -1 \\ 1 & -1 & 0 \\ 0 & 1 & -1 \end{bmatrix},$$

$$C' = \begin{bmatrix} 1 & 1 & 1 & -2 \\ 1 & 1 & -2 & 1 \\ 1 & -2 & 1 & 1 \end{bmatrix}.$$

Example 8. Consider 5-point cyclic convolution

$$c(x) \equiv g(x)h(x) \; mod \; (x^5 - 1).$$

Factorization (2) is

$$x^5 - 1 = (x - 1)(x^4 + x^3 + x^2 + x + 1).$$

Then

$$M_1 = \begin{bmatrix} 1 & 1 & 1 & 1 & 1 \end{bmatrix},$$

$$M_2 = \begin{bmatrix} 1 & 0 & 0 & 0 & -1 \\ 0 & 1 & 0 & 0 & -1 \\ 0 & 0 & 1 & 0 & -1 \\ 0 & 0 & 0 & 1 & -1 \end{bmatrix}.$$

Directly

$$A_1 = B_1 = C_1 = \begin{bmatrix} 1 \end{bmatrix}.$$

Multiplication $mod \; (x^4 + x^3 + x^2 + x + 1)$ can be computed by first computing 4×4 linear convolution by the algorithm designed in example 6. Using the notation of example 6,

$$A_2 = B_2 = A',$$

$$C_2 = C'.$$

Direct computation shows that

$$AM = \begin{bmatrix} 1 & 1 & 1 & 1 & 1 \\ 1 & 0 & 0 & 0 & -1 \\ 1 & 0 & -1 & 0 & 0 \\ 1 & -1 & -1 & 1 & 0 \\ 0 & 1 & 0 & -1 & 0 \\ 1 & -1 & 1 & -1 & 0 \\ 1 & 0 & -1 & 1 & -1 \\ 1 & -1 & 0 & 1 & -1 \\ 0 & 1 & -1 & 0 & 0 \\ 0 & 0 & 0 & 1 & -1 \end{bmatrix} = BM.$$

To complete the ingrediants needed for (18), we observe that the idempotents are

$$e_1(x) = \frac{1}{3}(1 + x + x^2 + x^3 + x^4),$$

$$e_2(x) = -\frac{1}{3}(-4 + x + x^2 + x^3 + x^4),$$

which can be used to compute

$$E_1 = \frac{1}{3}\begin{bmatrix} 1 & 1 & 1 & 1 & 1 \\ 1 & 1 & 1 & 1 & 1 \\ 1 & 1 & 1 & 1 & 1 \\ 1 & 1 & 1 & 1 & 1 \\ 1 & 1 & 1 & 1 & 1 \end{bmatrix},$$

$$E_2 = -\frac{1}{3}\begin{bmatrix} -4 & 1 & 1 & 1 & 1 \\ 1 & -4 & 1 & 1 & 1 \\ 1 & 1 & -4 & 1 & 1 \\ 1 & 1 & 1 & -4 & 1 \\ 1 & 1 & 1 & 1 & -4 \end{bmatrix}.$$

6.6.   Digital Filters

The bilinear algorithms computing convolution, developed from the CRT, have the form

$$\underline{s} = C((B\underline{h})(A\underline{g})), \tag{1}$$

where C is usually more complicated than A or B. For application to digital filtering, we typically have one of the inputs, say $\underline{h}$ fixed, at least over many occurences, and $\underline{g}$ varies. We will now discuss the concept of the *transpose* of (1) which permits $\underline{s}$ to be computed by the formula,

$$\underline{s} = (\tilde{B})^t(((\tilde{C})^t\underline{h})(A\underline{g})), \tag{2}$$

where $\tilde{B}$ is the matrix determined by reversing the columns of B and $\tilde{C}$ is the matrix determined by reversing the rows of C.

Since $\underline{h}$ can be viewed as fixed, the computation $(\tilde{C})^t \cdot \underline{h}$ can be made once and for all. This precomputation stage, once made, does not enter into the overall efficiency of the algorithm which now depends on the matrices A and B. In the examples of section 5, the entries of A and B were always 0, 1 and -1, and that makes very obvious the advantage of the pre-computation $(\tilde{C})^t\underline{h}$.

In [3], the implications of this discussion to the stability of the computation were studied.

We turn now to a proof of (2). The result depends on the following observation about Toeplitz matrices.

Let T be a Toeplitz matrix which admits the factorization

$$T = CDB, \tag{3}$$

and let R denote a matrix of the same size of T given by

$$R = \begin{bmatrix} 0 & 0 & \cdots & 0 & 1 \\ 0 & 0 & \cdots & 1 & 0 \\ \cdots & & & & \\ 0 & 1 & \cdots & 0 & 0 \\ 1 & 0 & \cdots & 0 & 0 \end{bmatrix}, \tag{4}$$

then

$$T^t = RTR = (RC)D(BR) = \tilde{C}D\tilde{B}, \tag{5}$$

which proves that

$$T = (\tilde{B})^t D^t (\tilde{C})^t. \tag{6}$$

Consider now

$$s(x) = g(x)h(x) \ mod \ (x^N - 1).$$

We can write

$$\underline{s} = C(\underline{h})\underline{g},$$

where $C(\underline{h})$ is the circulant and, hence, the Toeplitz matrix with the vector $\underline{h}$ as its first column. Suppose we have a bilinear algorithm

$$\underline{s} = C((A\underline{g})(B\underline{h})) \tag{7}$$

computing $\underline{s}$. Let D be the diagonal matrix

$$D = diag. \quad (A\underline{g}), \qquad (8)$$

Then, we write

$$\underline{g} = CDB\underline{h}, \qquad (9)$$

and

$$C(\underline{h}) = CDB. \qquad (10)$$

By (6),

$$C(\underline{h}) = (\tilde{B})^t D(\tilde{C})^t, \qquad (11)$$

from which it follows that

$$\underline{s} = (\tilde{B})^t((A\underline{g})((\tilde{C})^t\underline{h})). \qquad (12)$$

We have additional results which can be proved from (6). For example, if $m(x) = x^4 - \xi$, where ξ is a constant, then

$$s(x) \equiv g(x)h(x) \, mod \, m(x) \qquad (13)$$

can be computed by

$$\underline{s} = T_\xi(\underline{h})\underline{g}, \qquad (14)$$

where $T_\xi(\underline{h})$ is the Toeplitz matrix

$$T_\xi(\underline{h}) = \begin{bmatrix} h_0 & \xi h_{N-1} & \cdots & \xi h_1 \\ h_1 & h_0 & \cdots & \xi h_2 \\ \cdots & & & \\ h_{N-1} & h_{N-2} & \cdots & h_0 \end{bmatrix}. \qquad (15)$$

Arguing as above, if $\underline{s} = C((A\underline{g})(B\underline{h}))$ is a bilinear algorithm, we have

$$\underline{s} = (\tilde{B})^t((A\underline{g})((\tilde{C})^t\underline{h})). \qquad (16)$$

Since

$$(A\underline{g})((\tilde{C})^t\underline{h})$$

is componentwise multiplication, the order can be changed, then

$$\underline{s} = (\tilde{B})^t(((\tilde{C})^t\underline{h})(A\underline{g})), \tag{17}$$

where vector $\underline{h}$ represents the system elements, and $\underline{g}$ is the input vector.

[References]

[1] Winograd, S. "Some Bilinear Forms Whose Multiplicative Complexity Depends on the Field of Constants", Math. Syst. Theor. 10(1977):pp. 169-180.

[2] Agarwal, R. C. and Cooley, J. W. "New Algorithms for Digital Convolution", IEEE Trans. on ASSP, 25(1977): pp. 392-410.

[3] Auslander, L., Cooley, J. W. and Silberger, A. J. "Number Stability of Fast Convolution Algorithms for Digital Filtering", VLSI Signal Proc., IEEE Press, (1984):pp. 172-213.

[4] Blahut, R. E. *Fast Algorithms for Digital Signal Processing*, Chapter 3 and 7. Addison-Wesley, 1985.

[5] Nussbaumer, H. J. *Fast Fourier Transform and Convolution Algorithms*, Second Edition, Chapter 3 and 6. Springer-Verlag, 1981.

[6] Burrus, C.S. and Parks, T.W. *DFT/FFT and Convolution Algorithms*, New York: John Wiley and Sons, 1985.

[7] Oppenheim, A.V. and Schafer, R.W. *Digital Signal Processing*, Englewood Cliffs, NJ: Prentice-Hall, 1975.

Problems

1. For two vectors $\underline{h} = [2, 3, 4, 5]$ and $\underline{g} = [6, 7, 8, 1]$, compute their linear convolution by

    a. Convolution summation.

    b. Polynomial multiplication.

    c. Matrix multiplication.

2. For two vectors $\underline{a} = [2, 3, 4, 5]$ and $\underline{b} = [6, 7, 8, 1]$, compute their cyclic convolution by

    a. Convolution summation.

    b. Polynomial multiplication.

    c. Matrix multiplication.

3. For $N = 5$ and $N = 6$, write the cyclic shift matrix S_5 and S_6, Prove that

$$S_5^5 = I_5, \qquad S_6^6 = I_6$$

4. For $\underline{a} = [1, 2, 3, 4, 5]$, write the circulant matrix $C(\underline{a})$, and represent $C(\underline{a})$ by the cyclic shift matrix S_5.

5. Compute the 4-point cyclic convolution by convolution theorem of the problem 2, show that the results are the same as the direct computation.

6. Diagonalize the matrix A,z

$$A = \begin{bmatrix} 4 & 8 & 1 & 2 \\ 2 & 4 & 8 & 1 \\ 1 & 2 & 4 & 8 \\ 8 & 1 & 2 & 4 \end{bmatrix}.$$

7. Show that $F(5)S_5 = D_5 F(5)$, where D_5 is a diagonal matrix, and give the diagonal matrix.

8. Take $m(x) = x(x - 1)(x + 1)(x + 2)(x - 2)(x - \infty)$, derive a complete Cook-Toom algorithm for 3 by 4 linear convolution.

9. Find the arithmetic counts of Problem 8.

Chapter 7

AGARWAL-COOLEY CONVOLUTION ALGORITHM

The cyclic convolution algorithms of chapter 6 are efficient for special small block lengths, but as the size of the block length increases, other methods are required. First as discussed in chapter 6, these algorithms keep the number of required multiplications small, but can require many additions. Also, each size requires a different algorithm. There is no uniform structure that can be repeatedly called upon. In this chapter, a technique similar to the Good-Thomas PFA will be developed to decompose a large size cyclic convolution into several small size cyclic convolutions which in turn can be evaluated using the Winograd cyclic convolution algorithm. These ideas were introduced by Agarwal and Cooley [1] in 1977. As in the Good-Thomas PFA, the CRT is used to define an indexing of data. This indexing changes a one-dimensional cyclic convolution into a two-dimensional cyclic convolution. We will see how to compute a two-dimensional cyclic convolution by 'nesting' a fast algorithm for one-dimensional cyclic convolution inside another fast algorithm for one-dimensional cyclic convolution. There are several two-dimensional cyclic convolution algorithms which although important will not be discussed. These can be found in [2].

7.1. Two-dimensional Cyclic Convolution

Consider two $M \times N$ matrices

$$\underline{g} = [g(m,n)]_{0 \leq m < M, \ 0 \leq n < N}, \tag{1}$$

$$\underline{h} = [h(m,n)]_{0 \leq m < M, \ 0 \leq n < N}. \tag{2}$$

We will define the *two-dimensional cyclic convolution*

$$\underline{s} = \underline{h} * \underline{g}. \tag{3}$$

Associate to $\underline{g}$ and $\underline{h}$, the polynomials in two variables,

$$G(x,y) = \sum_{m=0}^{M-1} \sum_{n=0}^{N-1} g(m,n)x^m y^n, \tag{4}$$

$$H(x,y) = \sum_{k=0}^{M-1} \sum_{l=0}^{N-1} h(k,l)x^k y^l. \tag{5}$$

Form the polynomial

$$S(x,y) = H(x,y)G(x,y) \bmod (x^M - 1) \bmod (y^N - 1), \tag{6}$$

by first forming the polynomial product

$$H(x,y)G(x,y), \tag{7}$$

then reducing $\bmod\,(x^M - 1)$ by setting $x^M = 1$ and in the same way, reducing $\bmod\,(y^N - 1)$ by setting $y^N = 1$. We can write

$$S(x,y) = \sum_{m=0}^{M-1} \sum_{n=0}^{N-1} s(m,n)x^m y^n. \tag{8}$$

We call the $M \times N$ matrix

$$\underline{s} = [s(m,n)]_{0 \le m < M,\ 0 \le n < N} \tag{9}$$

the *cyclic convolution* of $\underline{h}$ and $\underline{g}$.

We can compute $\underline{s}$ by the following *nesting procedure*. First by accumulating all the terms of (4), which have the same power of x, we can rewrite (4) as

$$G(x,y) = \sum_{m=0}^{M-1} g_m(y)x^m, \tag{10}$$

where

$$g_m(y) = \sum_{n=0}^{N-1} g(m,n)y^n, \quad 0 \le m < M. \tag{11}$$

In the same way, we can rewrite (5) and (8) as

$$H(x,y) = \sum_{m=0}^{M-1} h_m(y)x^m,\qquad (12)$$

$$S(x,y) = \sum_{m=0}^{M-1} s_m(y)x^m,\qquad (13)$$

then

$$s_l(y) \equiv \sum_{m=0}^{M-1} h_{l-m}(y)g_m(y)\ mod\ (y^N - 1),\quad 0 \le l < M,\qquad (14)$$

which can be viewed as *cyclic convolution mod M* where the data is no longer taken from the complex field but is taken from the ring $\mathbb{C}skip7pt[y]/(y^N - 1)$.

Main Idea   The cyclic convolution algorithms of chapter 6, designed for complex data, hold equally well for data taken from any ring containing the complex field. In particular, they can be used to compute (14). In this case, multiplication means multiplication in $\mathbb{C}skip7pt[y]/(y^N - 1)$ and addition means addition in $\mathbb{C}skip7pt[y]/(y^N - 1)$. The multiplication is cyclic convolution *mod N*.

Suppose cyclic convolution *mod M* is computed by an algorithm using $a(M)$ additions and $a(M)$ multiplications, with similar notation for cyclic convolution *mod N*. Then the M polynomials of Eq. (14)

$$s_l(y),\quad 0 \le l < M,\qquad (15)$$

are computed using $m(M)$ N-point cyclic convolutions and $a(M)$ additions in $\mathbb{C}skip7pt[y]/(y^N - 1)$. Since each N-point cyclic convolution is computed using $a(N)$ additions and $m(N)$ multiplications, we have that

$$m(M)m(N)\qquad (16)$$

multiplications and

$$N a(M) + m(M) a(N) \tag{17}$$

additions are needed to compute $\underline{s}$.

The order of the operations above can be interchanged reversing to roles of M and N on (16) and (17). This has no effect on multiplications but does effect additions.

We will now translate this discussion into matrix language. The M polynomial computations given in (14) can be rewritten as

$$
\begin{pmatrix} s_0(y) \\ s_1(y) \\ \vdots \\ s_{M-1}(y) \end{pmatrix} = \begin{pmatrix} h_0(y) & h_{M-1}(y) & \cdots & h_1(y) \\ h_1(y) & h_0(y) & \cdots & h_2(y) \\ & \cdots & & \\ h_{M-1}(y) & h_{M-2}(y) & \cdots & h_0(y) \end{pmatrix}
$$

$$
\times \begin{pmatrix} g_0(y) \\ g_1(y) \\ \vdots \\ g_{M-1}(y) \end{pmatrix} mod(y^N - 1). \tag{18}
$$

The matrix

$$
H(y) = \begin{pmatrix} h_0(y) & h_{M-1}(y) & \cdots & h_1(y) \\ h_1(y) & h_0(y) & \cdots & h_2(y) \\ & \cdots & & \\ h_{M-1}(y) & h_{M-2}(y) & \cdots & h_0(y) \end{pmatrix} \tag{19}
$$

is a circulant matrix having coefficients in $\mathbb{Q}skip7pt[y]/(y^N - 1)$. Set

$$
\underline{g}_m = \begin{pmatrix} g(m,0) \\ g(m,1) \\ \vdots \\ g(m, N-1) \end{pmatrix}, \qquad 0 \le m < M. \tag{20}
$$

and observe that $\underline{g}_m$ is the vector formed from the m-th row of the matrix g. In the same way, define $\underline{h}_l$, $0 \le l < M$ and $\underline{s}_k$, $0 \le k < M$.

Let H_l denote the circulant matrix having $\underline{h}_l$ as the 0-th column. We can rewrite (18) as

$$
\begin{pmatrix} \underline{s}_0 \\ \underline{s}_1 \\ \vdots \\ \underline{s}_{M-1} \end{pmatrix} = H \begin{pmatrix} \underline{g}_0 \\ \underline{g}_1 \\ \vdots \\ \underline{g}_{M-1} \end{pmatrix}, \tag{21}
$$

where H is the block-circulant matrix with circulant blocks:

$$
H = \begin{pmatrix} H_0 & H_{M-1} & \cdots & H_1 \\ H_1 & H_0 & \cdots & \cdot \\ \vdots & \vdots & \vdots & \vdots \\ H_{M-1} & \cdot & \cdots & \cdot \end{pmatrix}. \tag{22}
$$

Eq.(21) is the *matrix description of two-dimentional cyclic convolution*.

In chapter 6, bilinear cyclic convolution algorithms were designed as matrix factorizations of circulant matrices. We will extend these one-dimensional algorithms to the two-dimensional computation given by H. Matrices A and B define a *bilinear N-point cyclic convolution algorithm*, if for any $N \times N$ circulant matrix C, a diagonal matrix G can be found satisfying

$$
C = BGA, \tag{23}
$$

A class of algorithms of this kind have been given in chapter 5. In the convolution theorem, we have $A = B^{-1} = F(N)$. In the Winograd algorithms, the matrices A and B are matrices of small integers but are no longer square matrices.

First consider the special case

$$
H = C \otimes C', \tag{24}
$$

where C is an $M \times M$ circulant matrix and C' is an $N \times N$ circulant matrix. H has the form (22). Take bilinear algorithms computing M-point and N-point cyclic convolution

$$
C = BGA, \tag{25}
$$

$$C' = B'G'A', \tag{26}$$

where G and G' are diagonal matrices. Placing (25) and (26) in (24), we can write,

$$H = (B \otimes B')(G \otimes G')(A \otimes A'), \tag{27}$$

where $G \otimes G'$ is a diagonal matrix.

Consider, again, the matrix H given in (22). By (26), we can write

$$H_l = B'G_l'A', \qquad 0 \le l < M, \tag{28}$$

where the diagonal matrix G_l' is determined by H_l. Placing (28) in (23), we can rewrite H as

$$H = (I_M \otimes B')D'(I_M \otimes A'), \tag{29}$$

where D' is the block-circulant matrix having diagonal matrix blocks

$$D' = \begin{pmatrix} G_0' & G_{M-1}' & \cdots & G_1' \\ G_1' & G_0' & \cdots & \cdot \\ \vdots & \vdots & \vdots & \vdots \\ G_{M-1}' & \cdot & \cdots & \cdot \end{pmatrix}. \tag{30}$$

Suppose the size of each diagonal matrix G_l', $0 \le l < M$, is K. Then A' is an $K \times N$ matrix and B' is an $N \times K$ matrix.

The matrix

$$P(MK,K)D'P(MK,M) \tag{31}$$

is a block-diagonal matrix consisting of K $M \times M$ circulant blocks. By (23), we can write (31) as the matrix direct sum

$$\sum_{k=0}^{K-1} \oplus BG_k A, \tag{32}$$

where G_k, $0 \le k < K$, is a diagonal matrix. This implies that D' can be written as

$$(B \otimes I_K)D(A \otimes I_K) \tag{33}$$

for some diagonal matrix D. Placing (33) in (29),

$$H = (B \otimes B')D(A \otimes A'). \tag{34}$$

We see that the bilinear algorithms (25) and (26) can be used to compute the two-dimensional cyclic convolution given by H. In particular, the convolution theorem,

$$H = (F(M) \otimes F(N))^{-1}D(F(M) \otimes F(N)), \tag{35}$$

is the *two-dimensional convolution theorem*.

7.2. Agarwal-Cooley Algorithm

The CRT will be used to turn a one-dimensional N-point cyclic convolution where

$$N = N_1 N_2, \qquad (N_1, N_2) = 1, \tag{1}$$

into a two-dimensional $N_1 \times N_2$ cyclic convolution. By the results of section 1, we can then carry out the computation by nesting an N_1-point cyclic convolution algorithm inside an N_2-point cyclic convolution. Formula (1.33) is the explicit form of this nesting.

Choose idempotents e_1 and e_2 satisfying

$$e_1 \equiv 1 \ modN_1, \qquad e_2 \equiv 0 \ modN_2, \tag{2}$$

$$e_2 \equiv 0 \ modN_1, \qquad e_1 \equiv 1 \ modN_2. \tag{3}$$

Each n, $0 \le n < N$, can be uniquely written as

$$n \equiv n_1 e_1 + n_2 e_2 \ modN, \qquad 0 \le n_1 < N_1, 0 \le n_2 < N_2. \tag{4}$$

Consider the N-point cyclic convolution

$$\underline{s} = \underline{h} * \underline{g}, \tag{5}$$

which we can rewrite in the form

$$\underline{s} = H\underline{g}, \tag{6}$$

where H is the circulant matrix

$$H = \begin{pmatrix} h(0) & h(N-1) & \cdots & h(1) \\ h(1) & h(0) & \cdots & h(2) \\ \cdots & & & \\ h(N-1) & h(N-2) & \cdots & h(0) \end{pmatrix}. \tag{7}$$

We will show that a permutation matrix P can be found such that

$$P\underline{s} = (PHP^{-1})P\underline{g}, \tag{8}$$

where PHP^{-1} is a block-circulant matrix with circulant blocks. As a consequence, formula (8) computes $P\underline{s}$ as the two dimensional cyclic convolution in the sense of formula (1.22).

Example 1. Take $N = 6$ with $N_1 = 2$ and $N_2 = 3$. The idempotents are

$$e_1 = 3, \qquad e_2 = 4.$$

Consider 6-point cyclic convolution

$$\underline{s} = H\underline{g},$$

where H is a 6×6 circulant matrix. Define the permutation π of $Z/6$ by

$$\pi = (0, 4, 2;\ 3, 1, 5),$$

and denote by P the corresponding permutation matrix. Then

$$P = \begin{pmatrix} 1 & 0 & 0 & 0 & 0 & 0 \\ 0 & 0 & 0 & 0 & 1 & 0 \\ 0 & 0 & 1 & 0 & 0 & 0 \\ 0 & 0 & 0 & 1 & 0 & 0 \\ 0 & 1 & 0 & 0 & 0 & 0 \\ 0 & 0 & 0 & 0 & 0 & 1 \end{pmatrix}.$$

Direct computation shows that

$$PHP^{-1} = \begin{pmatrix} h(0) & h(2) & h(4) & h(3) & h(5) & h(1) \\ h(4) & h(0) & h(2) & h(1) & h(3) & h(5) \\ h(2) & h(4) & h(0) & h(5) & h(1) & h(3) \\ & & & & & \\ h(3) & h(5) & h(1) & h(0) & h(2) & h(4) \\ h(1) & h(3) & h(5) & h(4) & h(0) & h(2) \\ h(5) & h(1) & h(3) & h(2) & h(4) & h(0) \end{pmatrix},$$

which is a 2×2 block circulant matrix having 3×3 circulant blocks. The input and output vectors are

$$P\underline{g} = \begin{pmatrix} g(0) \\ g(4) \\ g(2) \\ g(3) \\ g(1) \\ g(5) \end{pmatrix}, \qquad P\underline{s} = \begin{pmatrix} s(0) \\ s(4) \\ s(2) \\ s(3) \\ s(1) \\ s(5) \end{pmatrix},$$

and

$$P\underline{s} = (PHP^{-1})P\underline{g},$$

is a two-dimensional 2×3 cyclic convolution.

Consider the general case $N = N_1 N_2$ where N_1 and N_2 are relatively prime. As in the Good-Thomas PFA, we define the permutation π of Z/N by the formula

$$\pi(n_2 + N_2 n_1) \equiv n_1 e_1 + n_2 e_2 \; modN, \qquad 0 \le n_1 < N_1, \; 0 \le n_2 < N_2. \tag{9}$$

Denote the corresponding permutation matrix by P. $P\underline{g}$ is formed by reading across the rows of the matrix

$$\begin{pmatrix} g(0) & g(e_2) & \cdots & g((N_2-1)e_2) \\ g(e_1) & g(e_1+e_2) & \cdots & g(e_1+(N_2-1)e_2) \\ \vdots & \vdots & \vdots & \vdots \\ g((N_1-1)e_1) & & \cdots & g((N_1-1)e_1+(N_2-1)e_2) \end{pmatrix}. \tag{10}$$

Write

$$PHP^{-1} = [\, h(l,k)\,] \qquad 0 \le l, k < N. \tag{11}$$

Then

$$h(n_2 + N_2 n_1, k_2 + N_2 k_1) = h((n_1 - k_1)e_1 + (n_2 - k_2)e_2), \quad (12)$$

where

$$0 \le n_1, k_1 < N_1, \qquad 0 \le n_2, k_2 < N_2.$$

From (12), we have that PHP^{-1} is a $N_1 \times N_1$ block circulant matrix having $N_2 \times N_2$ circulant blocks

$$PHP^{-1} = \begin{pmatrix} H_0 & H_{N_1-1} & \cdots & H_1 \\ H_1 & H_0 & \cdots & \cdot \\ \vdots & \vdots & \vdots & \vdots \\ H_{N_1-1} & \cdot & \cdots & H_0 \end{pmatrix}, \quad (13)$$

where H_l is the circulant matrix having 0-th column

$$\begin{pmatrix} h(le_1) \\ h(le_1 + e_2) \\ \vdots \\ h(le_1 + (N_2 - 1)e_2) \end{pmatrix}. \quad (14)$$

As a result

$$P\underline{s} = (PHP^{-1})P\underline{g}, \quad (15)$$

is a two-dimensional $N_1 \times N_2$ cyclic convolution.

Fast algorithms computing two-dimensional cyclic convolution (see (1.32) and (1.34)) can now be applied to compute (15) and in this way, the N-point cyclic convolution. If N_1-point cyclic convolution is computed using $a(N_1)$ additions and $m(N_1)$ multiplications, then

$$m(N_1)m(N_2) \quad (16)$$

multiplications are needed to compute the N-point cyclic convolution $\underline{s}$ by (15) and

$$N_2 a(N_1) + m(N_1)a(N_2) \quad (17)$$

additions are requered.

[References]

[1] Agarwal, R. C. and Cooley, J. W. "New Algorithms for Digital Convolution", IEEE Trans. ASSP-25 (1977):pp.392-410.

[2] Blahut, R. E. *Fast Algorithms for Digital Signal Processing*, Chapter 7. Addison-Wesley, 1985.

[3] Nussbaumer, H. J. *Fast Fourier Transform and Convolution Algorithms*, Second Edition, Chapter 6, Springer-Verlag, 1981.

[4] Arambepola, B. and Rayner, P. J. "Efficient Transforms for Multidimensional Convolutions", Electron. Lett. 15, (1979):pp.189-190.

Problems

1. Derive a 3-point Winograd cyclic convolution algorithm.

2. Derive a 5-point Winograd cyclic convolution algorithm.

3. Derive a 15-point Agarwal-Cooley convolution algorithm using the results of Problem 1 and 2.

4. Derive a 4-point Winograd cyclic convolution algorithm.

5. Derive a 12-point Agarwal-Cooley convolution algorithm using the results of Problem 1 and 4.

6. Write a 2×2 cyclic convolution algorithm

$$S(x,y) = H(x,y)G(x,y), \qquad (mod\ x^2 - 1)(mod\ y^2 - 1).$$

7. Write a 2×2 polynomial product

$$S(x,y) = H(x,y)G(x,y), \qquad (mod\ x^2 + 1)(mod\ y^2 + 1).$$

Chapter 8

INTRODUCTION TO MULTIPLICATIVE

FOURIER TRANSFORM ALGORITHM (MFTA)

The Cooley-Tukey FFT algorithm and its variants depend upon the existence of non-trivial divisors of the transform size N. These algorithms are called *additive algorithms* since they rely on the subgroups of the additive group structure of the indexing set. A second approach to the design of FT algorithms depends on the multiplicative structure of the indexing set. We appealed to the multiplicative structure previously, in chapter 5, in the derivation of the Good-Thomas PFA.

In the following chapters, a more extensive application of multiplicative structure will be required. The first breakthrough is due to Rader [1] in 1968, who observed that a p-point Fourier transform could be computed by a $(p-1)$ point cyclic convolution. Winograd [2-3] generalized Rader's results to include the case of transform size $N = p^m$, p a prime. Combined with Winograd's cyclic convolution algorithms, these methods lead to the *Winograd Small FT Algorithm* which we will derive in detail in chapter 9.

In table 1 and 2, taken from Temperton [4], we compare the number of real additions and real multiplications required by conventional methods and by the Winograd methods. For the transform sizes included in table 1 and 2, the Winograd's algorithm requires substantially fewer multiplications at the cost of a few extra additions. However, as the transform size increases, although the Winograd algorithm continues to maintain its advantage in multiplications, the price in additions becomes more costly. This is to be expected since these algorithms depend on cyclic convolution algorithms. Standing alone, the Winograd small Fourier transform algorithms are practical only for selected small size transforms. In tandem with the Good-Thomas PFA, the Winograd algorithms can be effectively used to handle medium and some large size transforms. *The Winograd Large FT algorithm* [5] is based on a method of nesting the WSFTA's in the Good-Thomas PFA. This results in algorithms which minimize multiplications. This nesting technique can be described using tensor product formulation. Suppose $N = RS$, where R and S are relatively prime. By the Good-Thomas algorithm

$$F(N) = P(F(R) \otimes F(S))Q, \tag{1}$$

where P and Q are permutation matrices defined by Good-Thomas method. A WSFTA for an R-point FFT has the form

$$F(R) = C_1 B_1 A_1. \tag{2}$$

In the cases under consideration, A_1 and C_1 are matrices of zeroes and ones and B_1 is a diagonal matrix whose entries are either real or purely imaginary. In the same way, we can write

$$F(S) = C_2 B_2 A_2. \tag{3}$$

The dimension of B_1 in (2) is in general greater than R and consequently A_1 and C_1 are not square matrices. Using the tensor product formula

$$(A \otimes B)(C \otimes D) = (AC) \otimes (BD), \tag{4}$$

we can place (2) and (3) in (1) and write the N-point Fourier transform matrix as

$$F(N) = PCBAQ, \tag{5}$$

where C and A are matrices of zeroes and ones:

$$C = C_1 \otimes C_2, \tag{6}$$

$$A = A_1 \otimes A_2, \tag{7}$$

and B is a diagonal matrix with real or purely imaginary entries on its diagonal

$$B = B_1 \otimes B_2. \tag{8}$$

The number of multiplications required to compute R-point FFT by (2) is the dimension $m(R)$ of the diagonal matrix B_1. It follows that the number of multiplication required to compute N-point FFT is

$$m(N) = m(R)m(S), \tag{9}$$

the dimension of the matrix B.

If $a(R)$ denotes the number of additions required to compute R-point FT by (2), then the number of additions required to compute the N-point FT by (5) is

$$a(N) = Ra(S) + m(S)a(R). \tag{10}$$

Kolba and Parks [6] implemented the Good-Thomas algorithm by direct computation of each small FFT using the Winograd FFT without nesting. In this case, the number of multiplications and additions are given respectively by

$$m(N) = Sm(R) + Rm(S). \tag{11}$$

$$a(N) = Sa(R) + Ra(S). \tag{12}$$

In general

$$R < m(R) \quad \text{and} \quad S < m(S), \tag{13}$$

which imply that the Kolba-Parks approach has the advantage when it comes to additions. However, in most cases, Winograd's approach has the advantage when measured by multiplications. Table 3–6, also taken from Temperton[4], compare the conventional approach, the Good-Thomas approach where conventional methods are used on the factors, the Kolba-Parks approach and the Winograd approach. As can be seen from table 4-6, the Winograd's technique offers substantial savings in multiplications relative to all other methods. However, it is the least efficient with respect to additions. In all cases, we see that additions dominate multiplications. Temperton in [7] argues for the Good-Thomas approach with conventional computation on the factors on computers such as CRAY where additions and multiplications are performed simultaneously. On these computers multiplications are "free" in the sense that they are carried out while the more numerous additions are being performed. Other implementation considerations are discussed in [8-9].

In the following chapters, we will present a class of algorithms [10-15], which combine features of all these algorithms. Tensor product rules are used throughout. The *fundamental factorization* has the form

$$F(N) = PCAP^{-1}, \qquad (14)$$

where P is a permutation matrix, A is a *pre-additions matrix* with all of its entries being 0, 1 or -1 and C is a block-diagonal matrix having skew-circulant blocks (*rotated Winograd cores*) and tensor product of such blocks.

We have implemented these algorithms and their variants on the Micro VAX II. For a large collection of transform sizes, these algorithms require roughly half the run-time of comparable transform sizes implemented by programs coming from Digital's Scientific Library (LabStar Version 1.1). We will see this in table 7 and 8.

Although the fundamental factorization given by (14) and its variants are highly structured and uniform, a direct attack on programming of matrix A is complicated. There is no apparent looping structure. However, as discussed in [14], the programming effort can be greatly simplified and automated by the use of macros and production rules which take advantage of the "local" structure of the pre-additions.

For all the transform size N's listed in tables 7 and 8, we used the Variant 1 form of factorization (14), as discribed in the following chapters. The main programs are written in Fortran and they call the small DFT subroutines written in assembly.

From tables 7 and 8, we can see that many transform sizes not included in Lab Star have been programmed. Timings for most sizes are significantly better than the nearest point Cooley-Tukey algorithm.

Tables of Arithmetic Counts and Timing

Table 1. Conventional Methods

Sizes	R.A.	R.M.
2	4	0
3	12	4
4	16	0
5	32	12
7	60	36
8	52	4
9	80	40
16	144	24

Table 2. Winograd Methods

Sizes	R.A.	R.M.
2	4	0
3	12	4
4	16	0
5	34	10
7	72	16
8	52	4
9	88	20
16	148	20

Table 3. Conventional Methods

Sizes	R.A.	R.M.
105	2272	1492
108	2018	1012
112	2162	1188
120	2302	1116
126	2684	1672
128	2242	900
240	5322	2708
252	5954	2500
256	5122	2050
315	8492	5728
320	7202	3396

Table 4. Good-Thomas Methods

Sizes	R.A.	R.M.
105	1992	932
112	1968	744
120	2028	508
126	2452	1208
240	4656	1256
252	5408	2416
315	7516	3776

Table 5. Kolba-Parks Methods

Sizes	R.A.	R.M.
105	2214	590
112	2188	396
120	2076	460
126	2780	568
240	4812	1100
252	6064	1136
315	8462	2050

Table 6. Winograd Methods

Sizes	R.A.	R.M.
105	2418	322
112	2332	308
120	2076	276
126	3068	392
240	5016	632
252	6640	784
315	10406	1186

R.A.– the number of real additions.

R.M.– the number of real multiplications.

Table 7. Timing Comparisons (pq and pqr cases)

Size	Factors	pq(pqr)	Dec.Labstar
8	2^3		1.49 ms.
15	3×5	1.13 ms.	
16	2^4		2.87 ms.
21	3×7	2.23 ms.	
32	2^5		5.78 ms.
33	3×11	4.35 ms.	
35	5×7	4.22 ms.	
39	3×13	4.78 ms.	
51	3×17	6.97 ms.	
64	2^6		12.49 ms.
93	3×31	16.03 ms.	
105	$3 \times 5 \times 7$	16.01 ms.	
128	2^7		27.80 ms.

Table 8. Timing Comparisons (4p ,4pq and p^2q cases)

Size	Factors	4p(4pq)	Dec.Labstar
8	2^3		1.49 ms.
12	4×3	0.415 ms.	
16	2^4		2.87 ms.
20	4×5	0.985 ms.	
28	4×7	2.92 ms.	
32	2^5		5.78 ms.
44	4×11	5.86 ms.	
45	$3^2 \times 5$	6.52 ms.	
52	4×13	6.53 ms.	
60	$4 \times 3 \times 5$	6.39 ms.	
64	2^6		12.49 ms.
68	4×17	9.27 ms.	
124	4×31	20.42 ms.	
128	2^7		27.80 ms.

where ms.$= 10^{-3}$ second.

[References]

[1] Rader, C. M. "Discrete Fourier Transforms When the Number of Data Samples Is Prime", Proc. IEEE 56 (1968):1107-1108.

[2] Winograd, S. "On Computing the Discrete Fourier Transform", *Proc. Nat. Acad. Sci. USA.*, vol. 73. no. 4,(April 1976):1005-1006.

[3] Winograd, S. "On Computing the Discrete Fourier Transform", *Math. of Computation*, Vol.32, No. 141, (Jan. 1978):pp 175-199.

[4] Temperton, C. "A Note on Prime Factor FFT Algorithms". *J. Comp. Phys.*, 52, (1983): 198-204.

[5] Blahut, R. E. *Fast Algorithms for Digital Signal Processing*, Chapter 8. Addison-Wesley, Reading, Mass., 1985.

[6] Kolba, D. P. and Parks, T. W. "Prime Factor FFT Algorithm Using High Speed Convolution", IEEE Trans. Acoust., Speech and Signal Proc. ASSP-25(1977):281-294.

[7] Temperton, C. "Implementation of Prime Factor FFT Algorithm on Cray-1", to be published.

[8] Agarwal, R.C. and Cooley, J. W. "Fourier Transform and Convolution Subroutines for the IBM 3090 Vector Facility", IBM J. Res. Devel., vol.30pp145-162. Mar., 1986.

[9] Agarwal, R.C. and Cooley, J.W. "Vectorized Mixed Radix Discrete Fourier Transform Algorithms", IEEE Proc. vol 75, no.9, Sep., 1987.

[10] Heideman, M. T.: *Multiplicative Complexity, Convolution, and the DFT*, Springer-Verlag 1988.

[11] Lu, Chao: *Fast Fourier Transform Algorithms For Special N's and The Implementations On VAX*. Ph.D. Dissertation. Jan. 1988, the City University of New York.

[12] Tolimieri, R. Lu, Chao and Johnson, W. R. : "Modified Winograd FFT Algorithm and Its Variants for Transform Size N=p^k and Their Implementations" accepted for publication by Advances in Applied Mathematics.

[13] Lu, Chao and Tolimieri, R.: "Extension of Winograd Multiplicative Algorithm to Transform Size N=p^2q, p^2qr and Their Implementation", Proceeding of ICASSP 89, Scotland, May 22-26.

[14] Gertner, Izidor: "A New Efficient Algorithm to Compute the Two- Dimensional Discrete Fourier Transform" IEEE Trans. on ASSP, Vol. 36, No. 7, July 1988.

[15] Johnson, R.W., Lu, Chao and Tolimieri, R.:*Fast Fourier Algorithms for the Size of Product of Distinct Primes and Implementations on VAX*. Submitted to IEEE Trans. Acout.,Speech, Signal Proc.

[16] Johnson, R. W., Lu, Chao and Tolimieri, R.: "Fast Fourier Algorithms for the Size of 4p and 4pq and Implementations on VAX". Submitted to IEEE Trans.Acout., Speech, Signal Proc.

Chapter 9

MFTA: THE PRIME CASE

9.1. The Field Z/p

For transform size p, p a prime, Rader introduced an approach to construct algorithms which depends on the multiplicative structure of indexing set. In fact, for a prime p, Z/p is a field and the unit group $U(p)$ is cyclic. Reordering input and output data corresponding to a generator of $U(p)$, the p-point FFT becomes essentially a $(p-1) \times (p-1)$ *skew-circulant* matrix. We require $2(p-1)$ additions to make this change. Rader computes this skew-circulant action by the convolution theorem which returns the computation to an FFT computation. Since the size $(p-1)$ is a composite number, the $(p-1)$-point FT can be handled by Cooley-Tukey FFT algorithms. The Winograd algorithm for small convolutions can also be applied to the skew-circulant action.

Take an odd prime p, the unit group $U(p)$ of Z/p is cyclic. We will use a generator z of $U(p)$ to reorder the indexing set.

<u>Example</u> 1. If $p = 3$, then $U(3)$ has the unique generator $z = 2$

<u>Example</u> 2. If $p = 5$, then $U(5)$ has two generators $z = 2$ and $z = 3$. Take $z = 2$, then we can order $U(5)$ by consecutive powers of 2

$$1, \ 2, \ 4, \ 3.$$

Analogously, if $z = 3$ is taken then we can order $U(5)$ by consecutive powers of 3

$$1, \ 3, \ 4, \ 2.$$

In the following table, we give generators z corresponding to several odd primes.

Table 1. Generator of $U(p)$

Size	Generator
3	2
5	2
7	3
11	2
13	2
17	6

Choose a generator z of $U(p)$. The order of $U(p)$ is $p - 1$. We will reorder the indexing set according to successive powers of the generator z as follows :

$$0, \ 1, \ z, \ z^2, \ \dots \ , \ z^{p-2}. \tag{1}$$

We call this ordering the *exponential ordering* associated to z. The relationship between the canonical ordering and the exponential ordering is given by the indexing set permutation

$$\pi = \begin{pmatrix} 0 & 1 & z & z^2 & \dots & z^{p-2} \end{pmatrix}, \tag{2}$$

which we call the *exponential permutation* associated to z.

Example 3. Relative to the generator 2 of $U(5)$, the exponential permutation is

$$\pi = \begin{pmatrix} 0 & 1 & 2 & 4 & 3 \end{pmatrix}.$$

Example 4. Relative to the generator 3 of $U(7)$, the exponential permutation is

$$\pi = \begin{pmatrix} 0 & 1 & 3 & 2 & 6 & 4 & 5 \end{pmatrix}.$$

A fact that will be useful in building algorithms is

$$z^{(p-1)/2} = -1. \tag{3}$$

Thus we can rewrite (2) as

$$0 \, , 1 \, , z \, , \, \ldots \, , \, z^{(p-1)/2} \, , \, -1 \, , \, -z \, , \, \ldots \, , \, -z^{(p-1)/2}. \qquad (4)$$

We assume throughout this section that a generator z of $U(p)$ has been specified, and that π is the exponential permutation relative to z. We will now design algorithms computing p-point Fourier transform based on reordering input and output data by the exponential ordering. In matrix formulation this amounts to permuting input and output data by the permutation matrix P corresponding to the exponential permutation π. Explicitly, P is the $p \times p$ permutation matrix defined by the formula

$$P\underline{x} = \underline{y}, \qquad (5)$$

where $y_0 = x_0$ and

$$y_k = x_{z^{k-1}}, \quad 1 \leq k < p. \qquad (6)$$

9.2. <u>The Fundamental Factorization</u> $N = p$

In chapter 2, we established a procedure for describing the matrix F_π defined by

$$F(p) = P^{-1} F_\pi P. \qquad (1)$$

First we compute the table of index products *mod p*.

	0	1	z		z^{p-2}
0	0	0	0	$\cdots$	0
1	0	1	z		z^{p-2}
z	0	z	z^2		1
$\vdots$	$\vdots$				
z^{p-2}	0	z^{p-2}	z		z^{p-3}

The matrix F_π is formed by replacing z^k in the table by v^{z^k}, $v = exp\,(2\pi i/p)$. It follows that

$$
F_\pi \;=\; \begin{bmatrix} 1 & 1 & \cdots & 1 \\ 1 & & & \\ \vdots & & C(p) & \\ 1 & & & \end{bmatrix}, \tag{2}
$$

where $C(p)$ is the $(p-1)$ by $(p-1)$ skew-circulant matrix

$$
C(p) \;=\; \big[\,v^{z^{l+k}}\,\big]_{0\leq l,k<p-1}. \tag{3}
$$

Observe that $C(p)$ depends upon the choice of generator z. We call $C(p)$ the *Winograd core* corresponding to the generator z of $U(p)$.

__Example__ 1. The Winograd core corresponding to the generator $z = 2$ of $U(3)$ is

$$
C(3) = \begin{bmatrix} v & v^2 \\ v^2 & v \end{bmatrix}, \quad v = exp(2\pi i/3).
$$

__Example__ 2. The Winograd core corresponding to the generator $z = 2$ of $U(5)$ is

$$
C(5) = \begin{bmatrix} v & v^2 & v^4 & v^3 \\ v^2 & v^4 & v^3 & v \\ v^4 & v^3 & v & v^2 \\ v^3 & v & v^2 & v^4 \end{bmatrix}, \quad v = exp(2\pi i/5).
$$

The factorization (1) of $F(p)$, where F_π is given by (2), was first obtained by Rader [1]. For any p-dimensional vector $\underline{x}$, denote by $\underline{x}'$ the $(p-1)$-dimensional vector formed by deleting the 0-th component from $\underline{x}$. The action of F_π can be computed by the following two formulas. If $\underline{y} = F_\pi \underline{x}$ then

$$
y_0 = \sum_{m=0}^{p-1} x_m, \tag{4}
$$

$$\underline{y}' = C(p)\underline{x}' + x_0 1_{p-1},\tag{5}$$

where 1_{p-1} is the $(p-1)$-dimensional vector of all ones.

Example 3. Corresponding to the generator $z = 2$ of $U(5)$, we can compute $\underline{y} = F_\pi \underline{x}$ by the formulas

$$y_0 = \sum_{m=0}^{4} x_m,$$

$$\begin{pmatrix} y_1 \\ y_2 \\ y_3 \\ y_4 \end{pmatrix} = \begin{pmatrix} v & v^2 & v^4 & v^3 \\ v^2 & v^4 & v^3 & v \\ v^4 & v^3 & v & v^2 \\ v^3 & v & v^2 & v^4 \end{pmatrix} \begin{pmatrix} x_1 \\ x_2 \\ x_3 \\ x_4 \end{pmatrix} + \begin{pmatrix} x_0 \\ x_0 \\ x_0 \\ x_0 \end{pmatrix}.$$

Up to permutation of input and output data by the permutation matrix P, the action of F_π computes the action of $F(p)$. Since the permutation matrix P has the form

$$P = \begin{pmatrix} 1 & 0 \\ 0 & Q \end{pmatrix},\tag{6}$$

where Q is a permutation matrix, we have

$$F(p) = \begin{pmatrix} 1 & 1 & \cdots & 1 \\ 1 & & & \\ \vdots & & Q^{-1}C(p)Q & \\ 1 & & & \end{pmatrix}.\tag{7}$$

The computation $\underline{y} = F_\pi \underline{x}$ can be carried out in several ways. Define the $p \times p$ matrix

$$A(p) = \begin{pmatrix} 1 & 1 & \cdots & 1 \\ -1 & & & \\ \vdots & & I_{p-1} & \\ -1 & & & \end{pmatrix}.\tag{8}$$

Observe that $A(p)$ does not depend on π.

Example 4.

$$A(3) = \begin{pmatrix} 1 & 1 & 1 \\ -1 & 1 & 0 \\ -1 & 0 & 1 \end{pmatrix}.$$

Example 5.

$$A(5) = \begin{pmatrix} 1 & 1 & 1 & 1 & 1 \\ -1 & 1 & 0 & 0 & 0 \\ -1 & 0 & 1 & 0 & 0 \\ -1 & 0 & 0 & 1 & 0 \\ -1 & 0 & 0 & 0 & 1 \end{pmatrix}.$$

Example 6. Consider the Winograd core $C(5)$ of example 3. Since

$$1 + v + v^2 + v^3 + v^4 = 0, \quad v = exp(2\pi i/5),$$

we have

$$C(5)1_4 = -1_4,$$

implying that

$$F_\pi = \begin{pmatrix} 1 & 0 & 0 & 0 & 0 \\ 0 & & & & \\ 0 & & C(5) & & \\ 0 & & & & \\ 0 & & & & \end{pmatrix} A(5).$$

Using the matrix direct sum notation, we can rewrite this as

$$F_\pi = (1 \oplus C(5))A(5),$$

which leads to the computation of $y = F_\pi x$ by the following sequence of steps:

1. **Compute**

$$a_0 = x_0 + x_1 + x_2 + \ldots + x_{p-1},$$

$$a_1 = x_1 - x_0,$$

$$a_2 = x_2 - x_0,$$

$$a_3 = x_3 - x_0,$$

$$\vdots$$

$$a_{p-1} = x_{p-1} - x_0.$$

2. **Compute**

$$y_0 = a_0,$$

$$\begin{pmatrix} y_1 \\ y_2 \\ \vdots \\ y_{p-1} \end{pmatrix} = C(p) \begin{pmatrix} a_1 \\ a_2 \\ \vdots \\ a_{p-1} \end{pmatrix}.$$

The results of example 6 hold in general. The main fact we need is

$$1 + \sum_{k=0}^{p-2} v^{z^k} = 0. \tag{9}$$

Then

$$C(p)1_{p-1} = -1_{p-1}, \tag{10}$$

which implies

$$F_\pi = (1 \oplus C(p))A(p), \tag{11}$$

In this factorization, $C(p)$ depends on the choice of the generator z of $U(p)$. The factorization given in (11) is called the *fundamental factorization*.

Formula (11) computes the action of F_π in two stages:

1. An "*additive stage*" described by the matrix $A(p)$.

2. A "*multiplicative stage*" computing the action of the skew-circulant Winograd core $C(p)$.

The additive stage requires $2(p-1)$ additions. In the next section, several implementations of the multiplicative stage will be given having various arithmetic counts. We have called this stage the multiplicative stage since by the convolution theorem, the skew-circulant matrix $C(p)$ can be diagonalized using $F(p-1)$.

Taking transpose on both sides of (11), we have

$$F_\pi = A^t(p)\big(1 \oplus C(p)\big). \tag{12}$$

the multiplicative stage now comes before the additive stage.

A second algorithm computing

$$\underline{y} = F_\pi \underline{x},$$

will now be given. Set

$$E(p) = 1_p^t \otimes 1_p, \tag{13}$$

the $p \times p$ matrix of all ones. Form the matrix

$$F_\pi - E(p),$$

which from (2) can be written as

$$F_\pi - E(p) = 0 \oplus W(p), \tag{14}$$

where W(p) is the $(p-1) \times (p-1)$ skew-circulant matrix given by

$$W(p) = C(p) - E(p-1). \tag{15}$$

The computation becomes

$$\underline{y} = (F_\pi - E(p))\underline{x} + E(p)\underline{x}. \tag{16}$$

We see that

$$E(p)\underline{x} = y_0 1_p \tag{17}$$

can be computed using $(p-1)$ additions. The computation is arranged in two stages.

1.
$$y_0 = \sum_{j=0}^{p-1} x_j$$

2.
$$\underline{y}' = W(p)\underline{x}' + y_0 1_{p-1}$$

As in the preceeding approach, we require $2(p-1)$ additions and the action of the $(p-1) \times (p-1)$ skew-circulant matrix $W(p)$.

Example 7. Corresponding to the generator $z = 2$ of $U(5)$,

$$W(p) = \begin{pmatrix} v-1 & v^2-1 & v^4-1 & v^3-1 \\ v^2-1 & v^4-1 & v^3-1 & v-1 \\ v^4-1 & v^3-1 & v-1 & v^2-1 \\ v^3-1 & v-1 & v^2-1 & v^4-1 \end{pmatrix}, \quad v = exp(2\pi i/5).$$

The computation of $\underline{y} = F_\pi \underline{x}$ can be carried out by

$$y_0 = \sum_{m=0}^{4} x_m,$$

$$\begin{pmatrix} y_1 - y_0 \\ y_2 - y_0 \\ y_3 - y_0 \\ y_4 - y_0 \end{pmatrix} = W(p) \begin{pmatrix} x_1 \\ x_2 \\ x_3 \\ x_4 \end{pmatrix}.$$

9.3. Rader's Algorithm

For a prime p, the factorization

$$F_\pi = (1 \oplus C(p))A(p), \tag{1}$$

decomposes the action of F_π into two stages.

Unless otherwise specified additions and multiplications mean complex additions and complex multiplications. Every addition is equivalent to two real additions. There are several ways of computing multiplications. We will assume that direct methods are used so that every multiplication requires two real additions and four real multiplications. In this and the next section, we will design variants of (1) that reduce multiplications or change the balance between the multiplications and additions.

By the convolution theorem, the matrix

$$D(p) = F(p')^{-1}C(p)F(p')^{-1}, \quad p' = p - 1, \tag{2}$$

is a diagonal matrix. In fact

$$D(p) = diag(\underline{d}), \tag{3}$$

where

$$\underline{d} = F(p')^{-1}\underline{C}_0. \tag{4}$$

$\underline{C}_0$ is the vector formed from the 0-th column of $C(p)$. From (2), we can write

$$C(p) = F(p')D(p)F(p'), \tag{5}$$

which can be used to compute the action of $C(p)$ in three stages.

From (5), we can write

$$1 \oplus C(p) = (1 \oplus F(p'))(1 \oplus D(p))(1 \oplus F(p')), \tag{6}$$

which leads to the *Rader factorization*

$$F_\pi = (1 \oplus F(p'))(1 \oplus D(p))(1 \oplus F(p'))A. \tag{7}$$

9.4. Reducing Additions

Diagonalizing the skew-circulant Winograd core $C(p)$ reduces the number of multiplications needed to carry out the computation. This is an important, but not the only, consideration even when small computers are used having fast adders and slow multipliers. Data flow can have a great effect on the actual time cost of carrying out a computation. However, measuring the efficiency of a given data flow is extremely machine dependent, and beyond the scope of this work.

On some larger machines, the speed of doing additions is nearly equal to that of multiplications. In this case, algorithms, striking a balance between number of additions and number of multiplications, should be most efficient. Several algorithms tending to reduce additions will be presented.

Consider first the Rader factorization

$$F_\pi = (1 \oplus F(p'))(1 \oplus D(p))(1 \oplus F(p'))A(p), p' = p - 1. \textbf{(Variant 1.)} \tag{1}$$

Interchanging the order of the factors will reduce the cost of additions with a slight increase in the cost of multiplications. From (1) we can write

$$F_\pi = (1 \oplus F(p'))(1 \oplus D(p))B(p)(1 \oplus F(p')), \quad \textbf{(Variant 2.)} \tag{2}$$

where

$$B(p) = (1 \oplus F(p'))A(p)(1 \oplus F(p')^{-1}). \tag{3}$$

In terms of arithmetic, the action of $A(p)$ is replaced by the action of $B(p)$. In addition, factorization (1) and (2) present two distinct data flows. In (2), $1 \oplus F(p')$ acts on both input and output. This can be a useful feature on certain machines.

We will describe $B(p)$. Let $\underline{e}$ be the vector of size $p' = p - 1$ defined by

$$\underline{e}^t = (1 \quad 0 \quad \dots \quad 0). \tag{4}$$

Since

$$A(p) = \begin{pmatrix} 1 & 1_{p'}^t \\ -1_{p'} & I_{p'} \end{pmatrix}, \tag{5}$$

direct computation using

$$F(p')\underline{e} = 1_{p'}, \tag{6}$$

shows that

$$B(p) = \begin{pmatrix} 1 & \underline{e}^t \\ -p'\underline{e} & I_{p'} \end{pmatrix}. \tag{7}$$

Example 1.

$$B(5) = \begin{bmatrix} 1 & 1 & 0 & 0 & 0 \\ -4 & 1 & 0 & 0 & 0 \\ 0 & 0 & 1 & 0 & 0 \\ 0 & 0 & 0 & 1 & 0 \\ 0 & 0 & 0 & 0 & 1 \end{bmatrix}.$$

Computing the action of $B(p)$ requires 2 additions and integer multiplications by $-p'$ which should be compared with the arithmetic of $A(p)$ given in table 1 of section 3. The additions have been dramatically reduced with one extra multiplication as the trade off. Notice another important factor that no matter how large p is, the arithmetic of $B(p)$ remains constant.

The second approach for reducing additions comes from the special form of the Winograd core $C(p)$.

Example 2. Take $p = 7$. The matrix $C(7)$ corresponding to the generator $z = 3$, has the form

$$C(7) = \begin{pmatrix} X(7) & X^*(7) \\ X^*(7) & X(7) \end{pmatrix},$$

where

$$X(7) = \begin{pmatrix} v & v^3 & v^2 \\ v^3 & v^2 & v \\ v^2 & v & v^3 \end{pmatrix}.$$

A straightfoward computation of the action of $C(7)$ requires 132 real additions and 144 real multiplications.

A *partial diagonalization method* will be used. Set

$$Y(7) = \frac{1}{2}((X(7) + X^*(7)) \oplus (X(7) - X^*(7))).$$

By direct computation, we have that

$$C(7) = (F(2) \otimes I_3)Y(7)(F(2) \otimes I_3).$$

Computing $C(7)$ by this formula results in savings in the cost of additions. First the matrix

$$X(7) + X^*(7),$$

has only real entries. Multiplication of a real number and a complex number requires no real additions and two real multiplications. It follows that the action of $X(7) + X^*(7)$ requires 18 real multiplications and 12 real additions. The matrix

$$X(7) - X^*(7),$$

has only purely imaginary entries. We assume that multiplication by i requires no addition or multiplication. The arithmetic of $X(7) - X^*(7)$ is equivalent to that of $X(7)+X^*(7)$. It follows that the action of $X(7) - X^*(7)$ requires 18 real multiplications and 12 additions. Since

$$1 \oplus C(7) = (1 \oplus (F(2) \otimes I_3))(1 \oplus Y(7))(1 \oplus (F(2) \otimes I_3)),$$

we have

$$F_\pi = (1 \oplus (F(2) \otimes I_3))(1 \oplus Y(7))(1 \oplus (F(2) \otimes I_3))A(7).$$

The computation of the action of F_π is decomposed into a pre-addition stage given by $A(7)$, a 2-point FFT stage given by $1 \oplus (F(2) \otimes I_3)$, an essentially real multiplication stage given by $1 \oplus Y(7)$ and a final 2-point FFT stage given by $1 \oplus (F(2) \otimes I_3)$. The arithmetic is summarized in table 4.1. Computing $C(7)$ by this method should be compared to the $p = 7$ case of table 3.3.

The general case follows in much the same way. From (2.3), $C(p)$ has the form

$$C(p) = \begin{pmatrix} X(p) & X^*(p) \\ X^*(p) & X(p) \end{pmatrix}. \qquad (8)$$

Set

$$Y(p) = \frac{1}{2}[(X(p) + X^*(p)) \oplus (X(p) - X^*(p))]$$

Direct computation shows that

$$C(p) = (F(2) \otimes I_{p'/2})Y(p)(F(2) \otimes I_{p'/2}). \tag{9}$$

Again, $X(p) + X^*(p)$ has only real entries and $X(p) - X^*(p)$ has only imaginary entries.

$$F_\pi = (1 \oplus (F(2) \otimes I_{p'/2}))(1 \oplus Y(p))(1 \oplus (F(2) \otimes I_{p'/2}))A(p).$$

$$\textbf{(Variant 3.)} \tag{10}$$

The preceeding two methods can be combined by writing

$$F_\pi = (1 \oplus (F(2) \otimes I_{p'/2}))(1 \oplus Y(p))B_1(p)(1 \oplus (F(2) \otimes I_{p'/2})),$$

$$\textbf{(Variant 4.)} \tag{11}$$

where

$$B_1(p) = (1 \oplus (F(2) \otimes I_{p'/2}))A(p)(1 \oplus (F(2) \otimes I_{p'/2})). \tag{12}$$

We will now describe $B_1(p)$. Let $\underline{f}$ be the p'-tuple vector whose first $p'/2$ entries are ones and last $p'/2$ entries are zeros,

$$\underline{f}^t = (1 \quad \cdots \quad 1 \quad 0 \quad \cdots \quad 0), \tag{13}$$

then

$$B_1(p) = \begin{pmatrix} 1 & \underline{f}^t \\ -2\underline{f} & I_{p'} \end{pmatrix}. \tag{14}$$

9.5. Winograd Small FFT Algorithm $N = p$

The action of the Winograd core $C(p)$ in the Rader factorization can also be computed by the Winograd small convolution algorithm.

Recall that the Winograd factorization for a skew-circulant matrix $C(p)$ has the form

$$C(p) = LGM, \tag{1}$$

where L and M are matrices of small integers and G is a diagonal matrix. The matrices L and M are generally not square matrices.

Example 1. Consider the 5-point FFT as given in example 2.2. Then,

$$F_\pi = \begin{pmatrix} 1 & 0 & 0 & 0 & 0 \\ 0 & & & & \\ 0 & & C(5) & & \\ 0 & & & & \\ 0 & & & & \end{pmatrix} A(5),$$

where

$$C(5) = \begin{pmatrix} v & v^2 & v^4 & v^3 \\ v^2 & v^4 & v^3 & v \\ v^4 & v^3 & v & v^2 \\ v^3 & v & v^2 & v^4 \end{pmatrix}, \quad v = exp(2\pi i/5).$$

Several Winograd factorizations of $C(5)$ have been derived in chapter 5. For example we have

$$C(5) = LGM,$$

where

$$L = \begin{pmatrix} 1 & 0 & -1 & 1 & 0 & 1 \\ 1 & -1 & 1 & 1 & -1 & 1 \\ -1 & 0 & 1 & 1 & 0 & 1 \\ -1 & 1 & -1 & 1 & -1 & 1 \end{pmatrix},$$

$$M = \begin{pmatrix} 1 & 0 & -1 & 0 \\ 1 & -1 & -1 & 1 \\ 0 & 1 & 0 & -1 \\ 1 & 0 & 1 & 0 \\ 1 & -1 & 1 & -1 \\ 0 & 1 & 0 & 1 \end{pmatrix},$$

and $G = diag(\underline{g})$ where

$$\underline{g} = \frac{1}{2} M \begin{pmatrix} v \\ v^3 \\ v^4 \\ v^2 \end{pmatrix} = \frac{1}{2} \begin{pmatrix} v - v^4 \\ v - v^4 - (v^3 - v^2) \\ v^3 - v^2 \\ v + v^4 \\ v + v^4 - (v^3 + v^2) \\ v^3 + v^2 \end{pmatrix}.$$

Then

$$F_\pi = \begin{pmatrix} 1 & 0 \\ 0 & L \end{pmatrix} \begin{pmatrix} 1 & 0 \\ 0 & G \end{pmatrix} \begin{pmatrix} 1 & 0 \\ 0 & M \end{pmatrix} A(5).$$

In general, if (1) is the Winograd factorization of the Winograd core $C(p)$, then we have the factorization

$$F_\pi = L' G' M', \tag{2}$$

where G' is the diagonal matrix

$$G' = 1 \oplus G, \tag{3}$$

and L' and M' are matrices of small integers

$$L' = 1 \oplus L, \tag{4}$$

$$M' = (1 \oplus M) A(p). \tag{5}$$

From (1),

$$F(p) = P^{-1} F_\pi P = P^{-1} L' G' M' P. \tag{6}$$

Set

$$A = M' P, \tag{7}$$

$$B = G', \tag{8}$$

and

$$C = P^{-1} L'. \tag{9}$$

The Winograd algorithm for p-point FT is given by

$$F(p) = CBA, \tag{10}$$

which has the same form as the Winograd Small FFT algorithm. This form was given in chapter 8. The computation of (10) can be carried out in three stages: the first stage is the pre-addition stage given by matrix A, the second stage is the multiplication stage given by the diagonal matrix B, and the last stage is the post-addition given by matrix C.

The Winograd algorithm increases the number of additions but greatly reduces the number of multiplications.

9.6. <u>Summary</u>

For a generator z of $U(p)$, we have defined the matrix F_π by the formula

$$F(p) = P^{-1} F_\pi P, \tag{1}$$

where π is the exponential permutation corresponding to z and P is the permutation matrix corresponding to π. F_π describes the p-point FFT on input and output data ordered exponentially by z. The fundamental factorization is given as

$$F_\pi = (1 \oplus C(p)) A(p), \tag{2}$$

where $A(p)$ is the pre-addition matrix

$$A(p) = \begin{pmatrix} 1 & 1^t_{p'} \\ -1_{p'} & I_{p'} \end{pmatrix}, \quad p' = p - 1, \tag{3}$$

and $C(p)$ is the $p' \times p'$ skew-circulant Winograd core.

$$C(p) = \begin{pmatrix} v & v^z & \cdots & v^{z^{p-2}} \\ v^z & v^{z^2} & \cdots & v \\ \vdots & & & \\ v^{z^{p-2}} & v & \cdots & v^{z^{p-3}} \end{pmatrix}, \quad v = exp(2\pi i/p). \tag{4}$$

Set $C = 1 \oplus C(p)$ and $A = A(p)$.

Variant 1.   (Rader Algorithm)

$$F_\pi = FDFA,\tag{5}$$

where $F = 1 \oplus F(p')$, $p' = p - 1$, and D is the diagonal matrix

$$D = 1 \oplus D(p),\tag{6}$$

with

$$D(p) = F(p')^{-1}C(p)F(p')^{-1}.\tag{7}$$

Variant 2.

$$F_\pi = FDB(p)F,\tag{8}$$

where

$$B(p) = FA(p)F^{-1} = \begin{pmatrix} 1 & \underline{e}^t \\ -p'\underline{e} & I_{p'} \end{pmatrix}, \quad p' = p - 1,\tag{9}$$

and $\underline{e}$ is the p'-tuple vector

$$\underline{e}^t = [1\ 0\ \ldots\ 0].\tag{10}$$

Write the skew-circulant matrix $C(p)$ in the form of

$$C(p) = \begin{pmatrix} X(p) & X^*(p) \\ X^*(p) & X(p) \end{pmatrix}.\tag{11}$$

Variant 3.

$$F_\pi = H(1 \oplus Y(p))HA(p),\tag{12}$$

where

$$H = 1 \oplus (F(2) \otimes I_{p'/2}), \quad p' = p - 1,\tag{13}$$

$$Y(p) = \frac{1}{2}((X(p) + X^*(p)) \oplus (X(p) - X^*(p))).\tag{14}$$

Variant 4.

$$F_\pi = H(1 \oplus Y(p))B_1(p)H,\tag{15}$$

where

$$B_1(p) = \begin{pmatrix} 1 & \underline{f}^t \\ -2\underline{f} & I_{p'} \end{pmatrix},$$
(16)

and $\underline{f}$ is the p'-tuple

$$\underline{f} = \begin{pmatrix} 1_{p'/2} \\ 0_{p'/2} \end{pmatrix}.$$
(17)

Winograd Algorithm By the Winograd Small Convolution algorithm, we have

$$C(p) = LGM,$$
(18)

where L and M are matrices of small integer and G is a diagonal matrix. We also have

$$F_\pi = (1 \oplus L)(1 \oplus G)(1 \oplus M)A(p).$$
(19)

The implementation of these algorithms has been carried out on the Micro VAX II. A major issue, apart from run-time, is simplicity of code generation. In particular, for the Micro VAX, variants 1 and 2 appear to have the best structure for programming. However, for implementation on computers such as the CRAY-XMP and IBM 3090, where multiplications are tied to additions, an arithmetically balanced algorithm is preferred. In this case, variant 4 is the best.

Tables of Arithmetic Counts

Table 3.1. $F_\pi = [1 \oplus C(p)]A(p)$ Direct Method

Factor	R.A.	R.M.
$A(p)$	4(p-1)	0
$C(p)$	2(p-1)(2p-3)	$4(p-1)^2$
F_π	$2(p-1)(2p-1)$	$4(p-1)^2$

Table 3.2. Arithmetic Count of $C(p)$: Direct Method

Factor	R.A.	R.M.
5	56	64
7	132	144
11	380	400
13	552	576
17	992	1024

Table 3.3. Arithmetic Count of $C(p)$: Convolution Theorem

Factor	R.A.	R.M.
5	38	12
7	82	36
11	202	76
13	214	76
17	326	100

Table 4.1. $C(7) = (F(2) \otimes I_3)Y(7)(F(2) \otimes I_3)$

Factor	R.A.	R.M.
$X + X^*$	12	18
$X - X^*$	12	18
Y	24	36
H	12	0
C	48	36

Table 4.2. $C(p)$ Variant 3.

Factor	R.A.	R.M.
$X(p) + X^*(p)$	$(p-1)(p-3)/2$	$(p-1)^2/2$
$X(p) - X^*(p)$	$(p-1)(p-3)/2$	$(p-1)^2/2$
$Y(p)$	$(p-1)(p-3)$	$(p-1)^2$
$F(2) \otimes I_{p'/2}$	$2(p-1)$	0
$C(p)$	$(p-1)(p+1)$	$(p-1)^2$
F_π	$(p-1)(p+5)$	$(p-1)^2$

Table 4.3. F_π Variant 4.

Factor	R.A.	R.M.
$B_1(p)$	2(p-1)	(p-1)
F_π	(p-1)(p+3)	p(p-1)

Table 4.4. F_π Variant 4

Factor	R.A.	R.M.
5	40	16
7	72	36
11	160	100
13	216	144
17	352	256
19	432	324

Table 4.5. F_π Variant 3

Factor	R.A	R.M
5	32	20
7	60	42
11	140	110
13	192	156
17	320	272
19	396	342

R.A. – the number of real additions.

R.M. – the number of real multiplications.

[References]

[1] Rader, C. M. "Discrete Fourier Transforms When the Number of Data Samples Is Prime", *Proc. IEEE* 56(1968):1107-1108.

[2] Winograd, S. "On Computing the Discrete Fourier Transform", *Proc. Nat. Acad. Sci. USA.*, vol. 73. no. 4, (April 1976):1005-1006.

[3] Winograd, S. "On Computing the Discrete Fourier Transform", *Math. of Computation*, Vol.32, No. 141, (Jan. 1978).

[4] Blahut, R.*Fast Algorithms for Digital Signal Processing* Addison-Wesley Pub. Co. 1885, Chapter 4.

[5] Heideman, M. T. *Multilicative Complexity, Convolution, and the DFT*, Chapter 5. Springer-Verlag, 1988.

[6] Johnson, R.W., Lu, Chao and Tolimieri, R.:"Fast Fourier Algorithms for the Size of Product of Distinct Primes and implementations on VAX". Submitted to *IEEE Trans. Acout.,Speech, Signal Proc.*

Problems

1. Show that 3, 2, 2 and 6 are generators of the unit group $U(7)$, $U(11)$, $U(13)$, and $U(17)$.

2. Find other generators for the unit group $U(7)$, $U(11)$, $U(13)$, and $U(17)$ that are different from the ones given in Problem 1.

3. Order the integer sets of $Z/7$, $Z/11$, $Z/13$ and $Z/17$ by the power of the generators given in Problem 1.

4. Write the Winograd core corresponding to the generator $z = 3$ for $U(7)$.

5. Write the Winograd core corresponding to the generator $z = 5$ for $U(7)$, observe the difference with Problem 4.

6. Since $F(7)$ can be written as

$$F(7) = \begin{bmatrix} 1 & 1 & \cdots & 1 \\ 1 & & & \\ \vdots & & Q^{-1}C(7)Q & \\ 1 & & & \end{bmatrix},$$

define the permutation matrix Q.

7. Show that the data given in table 3 of section 3 are correct.

8. Derive Rader algorithm for 11-point Fourier transform.

9. Computer the $C(11)$ Winograd core by $F(10)$, and $F(10)$ can be computed by Good-Thomas algorithm usibng $F(2)$ and $F(5)$, derive the Good-Thomas algorithm for $F(10)$.

10. If a given computer has the CPUTIME ratio as that one real multiplication equals ten real additions, what is the threshold size p, for which we choose Variant 2 to compute $F(p)$ instead of variant 1.

11. Prove the formulas given in Table 2 of section 4.

12. Discuss what kind of computer architecture is suitable for variant 4 algorithm.

13. What is the basic difference between the Winograd algorithm with the algorithms derived in sections 2, 3 and 4.

14. Give the arithmetic counts for the 5-point Winograd algorithm.

15. Derive the Winograd algorithm for $F(3)$.

Chapter 10

MFTA: PRODUCT OF TWO DISTINCT PRIMES

10.1. Basic Algebra

The results of chapter 9 will now be extended to the case of transform size N, N a product of two distinct primes. As mentioned in the general introduction to multiplicative FT algorithms, several approaches exist for combining small size FT algorithms into medium or large size FT algorithms by the Good-Thomas FT algorithms. Our approach emphasizes and is motivated by the results of chapter 9. By employing tensor product rules to a generalization of Rader's multiplicative FT algorithms, we derive the fundamental factorization

$$F_\pi = CA, \tag{1}$$

where C is a block-diagonal matrix having skew-circulant blocks (rotated Winograd cores) and tensor products of these skew-circulant blocks and A is a matrix of pre-additions, all of whose entries are 0, 1 or -1. Variants will then be derived.

Take $N = pq$, where p and q are distinct primes, and consider the ring Z/N. The unit group of Z/N

$$U(N) = \{a \in Z/N : (a, n) = 1\}, \tag{2}$$

is not a cyclic group. To determine the structure of $U(N)$, we will use the Chinese Remainder Theorem. First we can find elements e_1 and e_2 in Z/N satisfying

$$e_1 \equiv 1 \bmod p, \qquad e_1 \equiv 0 \bmod q, \tag{3}$$

$$e_2 \equiv 0 \bmod p, \qquad e_2 \equiv 1 \bmod q. \tag{4}$$

The set

$$\{e_1, e_2\} \tag{5}$$

is called a complete system of idempotents of Z/N. In chapter 5, in the derivation of the Good-Thomas Prime Factor FT algorithm, we defined the ring-isomorphism

$$\delta: Z/p \times Z/q \cong Z/N \tag{6}$$

by the formula

$$\delta(a_1, a_2) \equiv a_1 e_1 + a_2 e_2 \; mod \; N, \qquad 0 \le a_1 < p, 0 \le a_2 < q. \tag{7}$$

The ring direct product $Z/p \times Z/q$ is taken with respect to coordinatewise addition and multiplication. The ring-isomorphism δ restricts to a group-isomorphism

$$\delta: U(p) \times U(q) \cong U(N). \tag{8}$$

Since $U(p)$ is a cyclic group, when p is a prime, $U(N)$ is the direct product of cyclic groups of order $p - 1$ and $q - 1$. Choose generators z_1 of $U(p)$ and z_2 of $U(q)$. From (7), every element $a \in Z/N$ can be written uniquely as

$$a \equiv a_1 e_1 + a_2 e_2 \; mod \; N, \qquad 0 \le a_1 < p, 0 \le a_2 < q, \tag{9}$$

which using (8) implies that every $u \in U(N)$ can be written uniquely as

$$u \equiv z_1^l e_1 + z_2^k e_2 \; mod \; N, \qquad 0 \le l < p - 1, 0 \le k < q - 1. \tag{10}$$

We order $U(N)$ lexicographically in the parameters l and k in (10), taking k to be the fastest running parameter.

The idempotents satisfy the following properties :

$$e_1^2 \equiv e_1 \; mod \; N, \qquad e_2^2 \equiv e_2 \; mod \; N, \tag{11}$$

$$e_1 e_2 \equiv 0 \bmod N, \tag{12}$$

$$e_1 + e_2 \equiv 1 \bmod N. \tag{13}$$

An element $a \in Z/N$ that is not a unit, is called a *zero divisor*. The set

$$e_1 U(N) = \{z_1^l e_1 : 0 \leq l < p - 1\}, \tag{14}$$

consists of all elements in Z/N which are divisible by q but not p and the set

$$e_2 U(N) = \{z_2^k e_2 : 0 \leq k < q - 1\}, \tag{15}$$

consists of all elements in Z/N divisible by p but not q. In (14) and (15), we use the description of $U(N)$ given by (10) and the property $e_1 e_2 \equiv 0 \bmod N$. Order the sets $e_1 U(N)$ and $e_2 U(N)$ by the parameters l and k, respectively. Denote the unit group $U(N)$ by U, the set of zero divisors of Z/N is partitioned by the sets

$$0U, \qquad e_1 U, \qquad e_2 U. \tag{16}$$

Order the indexing set Z/N by the permutation

$$\pi = (0U, \ e_1 U, \ e_2 U, \ U), \tag{17}$$

and denote the corresponding $N \times N$ permutation matrix by P. It should be noted that π is not the same as the Good-Thomas permutation. The matrix

$$F_\pi = PF(N)P^{-1} \tag{18}$$

describes N-point FT where input and output data are ordered by π.

10.2. Transform Size $N = 15$

Take $p = 3$, $q = 5$ and $N = 15$. In this case, the idempotents are

$$e_1 = 10, \qquad e_2 = 6, \tag{1}$$

and every element $a \in Z/15$ can be written uniquely as

$$a \equiv a_1 \cdot 10 + a_2 \cdot 6 \bmod 15, \qquad 0 \le a_1 < 3, 0 \le a_2 < 5. \qquad (2)$$

Take generators $z_1 = 2$ of $U(3)$ and $z_2 = 2$ of $U(5)$. Every element $u \in U(15)$ can be written as

$$u = 2^l \cdot 10 + 2^k \cdot 6, \qquad 0 \le l < 2, \ 0 \le k < 4. \qquad (3)$$

$U(15)$ is ordered as follows :

$$1, \ 7, \ 4, \ 13, \ 11, \ 2, \ 14, \ 8. \qquad (4)$$

The set of zero divisors is partitioned into the sets

$$0 \cdot U(15) = \{0\}, \qquad (5)$$

$$10 \cdot U(15) = \{10, 5\}, \qquad (6)$$

$$6 \cdot U(15) = \{6, 12, 9, 3\}. \qquad (7)$$

Order the indexing set $Z/15$ by the permutation

$$\pi = (0; \ 10, 5; \ 6, 12, 9, 3; \ 1, 7, 4, 13, \ 11, 2, 14, 8) , \qquad (8)$$

The matrix M of index products mod 15 correspondong to π

$$M = [\pi(j)\pi(k)] , \qquad 0 \le j, k < 15. \qquad (9)$$

We now proceed to describe the matrix

$$F_\pi = \left[w^{\pi(j)\pi(k)} \right] , \qquad 0 \le j, k < 15, \ w = exp(2\pi i/15).$$

Set

$$u = w^5 = exp(2\pi i/3)$$

$$v = w^3 = exp(2\pi i/5), \qquad (10)$$

and set

$$C_3 = \begin{pmatrix} u^2 & u \\ u & u^2 \end{pmatrix}, \tag{11}$$

$$C_5 = \begin{pmatrix} v^2 & v^4 & v^3 & v \\ v^4 & v^3 & v & v^2 \\ v^3 & v & v^2 & v^4 \\ v & v^2 & v^4 & v^3 \end{pmatrix}. \tag{12}$$

The matrices C_3 and C_5 are *rotated Winograd cores*. C_3 is formed by by replacing u by u^2 in $C(3)$ and C_5 is formed by replacing v by v^2 in $C(5)$. Direct computation shows that the bottom right-hand 8×8 submatrix of F_π is the tensor product

$$C_3 \otimes C_5 = \begin{pmatrix} u^2 C_5 & u C_5 \\ u C_5 & u^2 C_5 \end{pmatrix}. \tag{13}$$

We can rewrite F_π as

$$F_\pi = \begin{pmatrix} 1 & 1_2^t & 1_4^t & 1_8^t \\ 1_2 & C_3 & E(2,4) & C_3 \otimes 1_4^t \\ 1_4 & E(4,2) & C_5 & 1_2^t \otimes C_5 \\ 1_8 & C_3 \otimes 1_4 & 1_2 \otimes C_5 & C_3 \otimes C_5 \end{pmatrix}, \tag{14}$$

where $E(M, N)$ is the $M \times N$ matrix of all ones. The highly structured form of F_π, especially the repetition of C_3 and C_5 throughout the matrix, results from controlling data flow by the idempotents.

10.3. The Fundamental Factorization $N = 15$

We will now derive for F_π a factorization of the form

$$F_\pi = CA, \tag{1}$$

where A is a matrix of additions and C is a block diagonal matrix having skew-circulant blocks. This will generalize the prime transform size factorization of the same form derived in the preceeding chapter. As in the preceeding chapter, factorization (1) will be a spring board for several collections of algorithms distinguished by arithmetic and data flow.

First

$$C_3 1_2 = -1_2, \tag{2}$$

$$C_5 1_4 = -1_4. \tag{3}$$

The tensor product formula

$$(A \otimes B)(C \otimes D) = (AC) \otimes (BD), \tag{4}$$

implies

$$C_3(I_2 \otimes 1_4^t) = C_3 \otimes 1_4^t, \tag{5}$$

$$C_5(1_2^t \otimes I_4) = 1_2^t \otimes C_5. \tag{6}$$

Using (2) and (3) along with

$$E(m,n) = 1_m \otimes 1_n^t, \tag{7}$$

we have

$$C_3 E(2,4) = -E(2,4), \tag{8}$$

$$C_5 E(4,2) = -E(4,2), \tag{9}$$

$$(C_3 \otimes C_5)1_8 = 1_8. \tag{10}$$

Placing these formulas in (2.15), we can write

$$F_\pi = CA, \tag{11}$$

where C is the block-diagonal matrix

$$C = 1 \oplus C_3 \oplus C_5 \oplus (C_3 \otimes C_5), \tag{12}$$

and

$$A = \begin{pmatrix} 1 & 1_2^t & 1_4^t & 1_8^t \\ -1_2 & I_2 & -E(2,4) & I_2 \otimes 1_4^t \\ -1_4 & -E(4,2) & I_4 & 1_2^t \otimes I_4 \\ 1_8 & -I_2 \otimes 1_4 & -1_2 \otimes I_4 & I_8 \end{pmatrix}. \tag{13}$$

We can relate the matrix A to the matrix of additions in chapter 9. Recall

$$A(p) = \begin{pmatrix} 1 & 1^t_{p'} \\ -1_{p'} & I_{p'} \end{pmatrix}, \qquad p' = (p-1), \tag{14}$$

We can rewrite A as

$$A = \begin{pmatrix} A(3) & A(3) \otimes 1^t_4 \\ -A(3) \otimes 1_4 & A(3) \otimes I_4 \end{pmatrix}. \tag{15}$$

Now let $Q_0 = P(12,4)$ as in chapter 2, and $Q = I_3 \oplus Q_0$. Straightforward computation shows that

$$Q_0(A(3) \otimes I_4)Q_0^{-1} = I_4 \otimes A(3), \tag{16}$$

$$(A(3) \otimes 1^t_4)Q_0^{-1} = 1^t_4 \otimes A(3), \tag{17}$$

$$Q_0(A(3) \otimes 1_4) = 1_4 \otimes A(3). \tag{18}$$

Placing these formulas in (15), we get

$$QAQ^{-1} = A(5) \otimes A(3), \tag{19}$$

and the factorization

$$F_\pi = CQ^{-1}(A(5) \otimes A(3))Q. \tag{20}$$

10.4. Variants $N = 15$

Variants of factorization

$$F_\pi = CA \tag{1}$$

will be designed in the spirit of chapter 9. First by the convolution theorem,

$$D_3 = F(2)^{-1}C_3 F(2)^{-1}, \tag{2}$$

$$D_5 = F(4)^{-1}C_5 F(4)^{-1}, \tag{3}$$

are diagonal matrices. Setting

$$F = 1 \oplus F(2) \oplus F(4) \oplus (F(2) \otimes F(4)), \qquad (4)$$

$$D = 1 \oplus D_3 \oplus D_5 \oplus (D_3 \otimes D_5), \qquad (5)$$

we have that D is a diagonal matrix and we can write

$$C = FDF, \qquad (6)$$

Placing (6) into (1), we get

$$F_\pi = FDFA, \qquad \text{(Variant 1.)}, \qquad (7)$$

which generalizes the Rader factorization for the prime case.

As in chapter 9, variants can be designed reducing additions. First rewrite (7) as

$$F_\pi = FDBF, \qquad \text{(Variant 2.)} \qquad (8)$$

where

$$B = FAF^{-1}. \qquad (9)$$

By interchanging the order of computation, the action of A is replaced by the action of B which, as we will now show, reduces the additions cost at the expense of a few rational multiplications. Define, as in chapter 9,

$$B(3) = \begin{pmatrix} 1 & 1 & 0 \\ -2 & 1 & 0 \\ 0 & 0 & 1 \end{pmatrix}, \qquad (10)$$

and

$$B(5) = \begin{pmatrix} 1 & 1 & 0 & 0 & 0 \\ -4 & 1 & 0 & 0 & 0 \\ 0 & 0 & 1 & 0 & 0 \\ 0 & 0 & 0 & 1 & 0 \\ 0 & 0 & 0 & 0 & 1 \end{pmatrix}. \qquad (11)$$

For the purpose of the discussion, set

$$X = 1 \oplus F(2), \tag{12}$$

and observe that

$$F = X \oplus (X \otimes F(4)). \tag{13}$$

A straightforward computation shows

$$XA(3) = B(3)X, \tag{14}$$

which is what we need to prove that

$$B = \begin{pmatrix} B(3) & B(3) \otimes \underline{e}^t \\ -4B(3) \otimes \underline{e} & B(3) \otimes I_4 \end{pmatrix}, \tag{15}$$

where

$$\underline{e}^t = \begin{bmatrix} 1 & 0 & 0 & 0 \end{bmatrix}. \tag{16}$$

We see that 40 real additions and 8 multiplications by small integers are needed to compute the action of B.

The action of B can also be computed, without changing the arithmetic, by the factorization

$$B = Q^{-1}(B(5) \otimes B(3))Q, \tag{17}$$

where $Q = I_3 \otimes P(12, 4)$.

The action of C requires complex multiplications which substantially increases the number of real additions. We apply the methods of chapter 9. to replace C by a matrix requiring only multiplications by real or purely imaginary numbers. C_3 and C_5 can be written in the form

$$C_3 = \begin{bmatrix} X_3 & X_3^* \\ X_3^* & X_3 \end{bmatrix}, \tag{18}$$

$$C_5 = \begin{bmatrix} X_5 & X_5^* \\ X_5^* & X_5 \end{bmatrix}. \tag{19}$$

Set

$$Y_3 = 1/2(X_3 + X_3^*) \oplus 1/2(X_3 - X_3^*), \tag{20}$$

$$Y_5 = 1/2(X_5 + X_5^*) \oplus 1/2(X_5 - X_5^*). \tag{21}$$

Then

$$C_3 = F(2)Y_3 F(2), \tag{22}$$

$$C_5 = (F(2) \otimes I_2)Y_5(F(2) \otimes I_2), \tag{23}$$

and

$$C_3 \otimes C_5 = (F(2) \otimes F(2) \otimes I_2)(Y_3 \otimes Y_5)(F(2) \otimes F(2) \otimes I_2). \tag{24}$$

Setting

$$H = 1 \oplus F(2) \oplus (F(2) \otimes I_2) \oplus (F(2) \otimes F(2) \otimes I_2), \tag{25}$$

$$Y = 1 \oplus Y_3 \oplus Y_5 \oplus (Y_3 \otimes Y_5), \tag{26}$$

we have

$$C = HYH. \tag{27}$$

Placing (27) into (1), we get

$$F_\pi = HYHA, \quad \text{(Variant 3.)} \tag{28}$$

Y contains all the multiplications. Reasoning as in the previous chapter, these are all real multiplications. Computing the action of F_π in this way requires 200 real additions and 68 real multiplications.

The cost of additions can be reduced by computing F_π by

$$F_\pi = HYB_1H, \text{(Variant 4.)} \tag{29}$$

where

$$B_1 = HAH^{-1}. \tag{30}$$

Using the notation of chapter 9, in particular

$$B_1(p) = \begin{pmatrix} 1 & \underline{f}^t(p) \\ -2\underline{f}(p) & I_{p'} \end{pmatrix}, \tag{31}$$

we can write

$$B_1 = \begin{pmatrix} B_1(3) & B_1(3) \otimes \underline{f}^t(5) \\ -2B_1(3) \otimes \underline{f}(5) & B_1(3) \otimes I_4 \end{pmatrix}, \tag{32}$$

where

$$\underline{f}^t(3) = [1, \, 0], \qquad \text{and} \qquad \underline{f}^t(5) = [1, \, 1, \, 0, \, 0].$$

As above, with $Q = I_3 \otimes P(12, 4)$,

$$B_1 = Q^{-1}(B_1(5) \otimes B_1(3))Q, \tag{33}$$

10.5. <u>General Case</u> $N = pq$

In this section, algorithms for $N = 15$ will be generalized to the case of $N = p.q$, where p and q are distinct primes.

Partition the indexing set Z/N by U-orbits, $U = U(N)$.

$$0 \cdot U = \{0\}, \tag{1}$$

$$e_1 \cdot U = \{z_1^k e_1 \mid 0 \le k < p - 1\}, \tag{2}$$

$$e_2 \cdot U = \{z_2^l e_2 \mid 0 \le l < q - 1\}, \tag{3}$$

$$U = \{z_1^k e_1 + z_2^l e_2 \mid 0 \le k < p - 1, 0 \le l < q - 1\}. \tag{4}$$

Order the indexing of input and output data by the permutation

$$\pi = (0; \, e_1 U; \, e_2 U; \, U). \tag{5}$$

To describe F_π we first form the matrix of products $mod \; N$,

$$M = [\, \pi(k) \; \pi(l) \,], \qquad 0 \le k, l < N, \tag{6}$$

It is convenient to breakup this matrix into submatrices given by the U-orbits:

	0	$e_1 U$	$e_2 U$	U
0				
$e_1 U$				
$e_2 U$				
U				

Consider the submatrix corresponding to the Cartesian product

$$e_1 U \times e_1 U. \tag{7}$$

A typical product in this submatrix is

$$(e_1 z_1^k)(e_1 z_1^r) = e_1^2 z_1^{k+l} \equiv e_1 z_1^{k+l} \ mod \ N, \tag{8}$$

from which it follows that this submatrix is

$$\left[e_1 z_1^{k+r} \right], \qquad 0 \leq k, r < p - 1. \tag{9}$$

Set

$$u_1 = v^{e_1}, \qquad v = exp(2\pi i/N), \tag{10}$$

and observe that u_1 is a primitive p-th root of unity. By (2), the submatrix of F_π corresponding to (7) is the skew-circulant matrix

$$C_p = \left[\ u_1^{z_1^{k+r}} \ \right], \qquad 0 \leq k, r < p - 1. \tag{11}$$

The powers of z_1 in (11) are taken $mod \ p$. The matrix C_p is a rotated Winograd core. It is formed by replacing $u = exp(2\pi i/p)$ in $C(p)$ by u_1. Any algorithm computing the action of $C(p)$ can easily be modified to be an algorithm computing the action of C_p.

In the same way, the submatrix of F_π corresponding to the Cartesian product $e_2 U \times e_2 U$ is the skew-circulant matrix (rotated Winograd core)

$$C_q = \left[\ u_2^{z_2^{l+s}} \ \right], \qquad 0 \leq l, s < q - 1, \tag{12}$$

where $u_2 = v^{e_2}$ is a primitive q-th root of unity.

Consider now the submatrix of F_π corresponding to the Cartesian product $U \times U$. A typical entry in this submatrix is given by raising v to the power

$$(z_1^k e_1 + z_2^l e_2)(z_1^r e_1 + z_2^s e_2) \equiv z_1^{k+r} e_1 + z_2^{l+s} e_2 \bmod N. \tag{13}$$

By (13), this becomes

$$u_1^{z_1^{k+r}} u_2^{z_2^{l+s}}. \tag{14}$$

Since l and s are the fastest running parameters, this submatrix can be decomposed into submatrices

$$u_1^{z_1^{k+r}} \cdot C_q, \qquad 0 \le k, r < p - 1, \tag{15}$$

it follows that the submatrix of F_π corresponding to $U \times U$ is the tensor product

$$C_p \otimes C_q. \tag{16}$$

Matrices (11),(12) and (16) take care of the diagonal blocks of F_π.

Similar arguments apply to the remaining submatrices. We now summarize the results in the following description of F_π,

$$F_\pi = \begin{pmatrix} 1 & 1_{p'}^t & 1_{q'}^t & 1_{r'}^t \\ 1_{p'} & C_p & E(p',q') & C_p \otimes 1_{q'}^t \\ 1_{q'} & E(q',p') & C_q & 1_{p'}^t \otimes C_q \\ 1_{r'} & C_p \otimes 1_{q'} & 1_{p'} \otimes C_q & C_p \otimes C_q \end{pmatrix}, \tag{17}$$

where $p' = p - 1$, $q' = q - 1$ and $r' = p'q'$. The hightly structured form of (17) is due in large part to the use of idempotents. If we set

$$F_\pi(p) = \begin{pmatrix} 1 & 1_{p'}^t \\ 1_{p'} & C_p \end{pmatrix}, \tag{18}$$

then

$$F_\pi = \begin{pmatrix} F_\pi(p) & F_\pi(p) \otimes 1_{q'}^t \\ F_\pi(p) \otimes 1_{q'} & F_\pi(p) \otimes C_q \end{pmatrix}, \tag{19}$$

which, by the permutation $P(pq,p)$ as defined in chapter 2, can be written as

$$F_\pi(p) \otimes F_\pi(q). \tag{20}$$

The form of $F_\pi(p)$ is the same as that of the matrix F_π, with a generator z of $U(p)$, defined in chapter 9. However, C_p is a rotated version of the Winograd core $C(p)$, where u is replaced by v^{e_1}.

10.6. Fundamental Factorization

The goal is to produce from the description of F_π given in (17) of the preceeding section a factorization of the form

$$F_\pi = CA, \tag{1}$$

where A is a matrix of additions and C is a block-diagonal matrix having skew-circulant blocks. The main ideas were given in the 15-point example. First since the sum of the m-th roots of unity equals zero for any integer $m > 1$, we have

$$C_p 1_{p'} = -1_{p'}, \qquad p' = p - 1, \tag{2}$$

$$C_q 1_{q'} = -1_{q'}, \qquad q' = q - 1. \tag{3}$$

Setting

$$C = 1 \oplus C_p \oplus C_q \oplus (C_p \otimes C_q), \tag{4}$$

a direct computation shows that the matrix A is given by

$$A = \begin{pmatrix} A(p) & A(p) \otimes 1^t_{q'} \\ -A(p) \otimes 1_{q'} & A(p) \otimes I_{q'} \end{pmatrix}, \tag{5}$$

Recall

$$A(m) = \begin{pmatrix} 1 & 1^t_{m'} \\ -1_{m'} & I_{m'} \end{pmatrix}, \qquad m' = m - 1. \tag{6}$$

The ring-structure has naturally pointed the way to the highly structured form of the factorization of F_π. As discussed in section 2, tensor product of the corresponding p-point and q-point algorithms directly leads to an arithmetically equivalent algorithm having a different data flow. This is a constant theme throughout this and the next section.

Using permutation matrices, we can implement A as a tensor product. To this end, setting

$$Q = I_p \oplus P(pq', q'), \tag{7}$$

and arguing as in chapter 2, we have that

$$A = Q^{-1}(A(q) \otimes A(p))Q. \tag{8}$$

It follows that

$$F_\pi = CQ^{-1}(A(q) \otimes A(p))Q. \tag{9}$$

From (9), we see that F_π can be built from the corresponding p-point and q-point algorithms designed in chapter 9. by tensor product. The arithmetic of direct computation of (1) or (9) can be described using results given in chapter 9. A general observation will be useful. If X is a $m \times m$ matrix, Y is a $n \times n$ matrix, and if an algorithm computes the actions of X and Y using $A(x)$ and $A(y)$ additions, respectively, then the action of the tensor product $X \otimes Y$ requires

$$n \cdot A(x) + m \cdot A(y) \tag{10}$$

additions.

10.7. Variants

The methods of chapter 9 will be applied to design several variants of the factorization

$$F_\pi = CA, \tag{1}$$

providing options for arithmetic and data flow which can be matched to a variety of computer architectures.

Variant 1. By the Convolution theorem

$$D_p = F(p')^{-1} C_p F(p')^{-1}, \qquad p' = p - 1, \tag{2}$$

and

$$D_q = F(q')^{-1} C_q F(q')^{-1}, \qquad q' = q - 1, \tag{3}$$

are diagonal matrices. Then

$$D = 1 \oplus D_p \oplus D_q \oplus (D_p \otimes D_q), \tag{4}$$

is a diagonal matrix and we can write

$$C = FDF, \tag{5}$$

where

$$F = 1 \oplus F(p') \oplus F(q') \oplus (F(p') \otimes F(q')). \tag{6}$$

This results in the factorization

$$F_\pi = FDFA. \tag{7}$$

Variant 2. The cost of additions in (7) can be reduced by interchanging the order of the operations. Write

$$F_\pi = FD(FAF^{-1})F \tag{8}$$

and set

$$B = FAF^{-1}. \tag{9}$$

Arithmetically, the action of A is replaced by the action B, which we will see requires fewer additions. To see this, we need to describe B. Let $\underline{e}(p)$ be p'-tuble whose 0-th entry is one and all other entries are zero. Recall the definition

$$B(p) = \begin{pmatrix} 1 & \underline{e}(p) \\ -p'\underline{e}(p) & I_{p'} \end{pmatrix}. \tag{10}$$

Direct computation shows that

$$(1 \oplus F(p'))A(p) = B(p)(1 \oplus F(p')),\qquad(11)$$

which, along with (7.6), is what we need to show that

$$B = \begin{pmatrix} B(p) & B(p) \otimes \underline{e}^t(q) \\ -p'B(p) \otimes \underline{e}(q) & B(p) \otimes I_{q'} \end{pmatrix}.\qquad(12)$$

Arguing as in the preceeding section, with

$$Q = I_p \otimes P(pq',q'),\qquad(13)$$

we have

$$B = Q^{-1}(B(q) \otimes B(p))Q,\qquad(14)$$

and

$$F_\pi = FDQ^{-1}(B(q) \otimes B(p))QF.\qquad(15)$$

The arithmetic of B is given as follows:

$$4(p+q)\qquad \text{R.A.}\qquad(16)$$

$$\{p+q\}\qquad \text{R.M.}\qquad(17)$$

In brackets, we have placed the number of multiplications by integers.

Variant 3. To reduce the cost of additions required to perform the complex multiplications coming from the action of C, we note that C_p has the form

$$C_p = \begin{pmatrix} X_p & X_p^* \\ X_p^* & X_p \end{pmatrix}.\qquad(18)$$

Direct computation shows that

$$C_p = (F(2) \otimes I_{p'/2})Y_p(F(2) \otimes I_{p'/2}),\qquad(19)$$

where

$$Y_p = 1/2(X_p + X_p^*) \oplus 1/2(X_p - X_p^*). \tag{20}$$

Arguing as before with

$$Y = 1 \oplus Y_p \oplus Y_q \oplus (Y_p \otimes Y_q), \tag{21}$$

$$H = 1 \oplus (F(2) \otimes I_{p'/2}) \oplus (F(2) \otimes I_{q'/2})$$

$$\oplus (F(2) \otimes I_{p'/2} \otimes F(2) \otimes I_{q'/2}), \tag{22}$$

we can write

$$C = HYH. \tag{23}$$

The next table uses the fact that $1/2(X_p + X_p^*)$ has only real entries and $1/2(X_p - X_p^*)$ has only purely imaginary entries.

Variant 4. From the factorization

$$F_\pi = HYHA, \tag{24}$$

we can reduce cost of additions by writing

$$F_\pi = HYB_1H, \tag{25}$$

where

$$B_1 = HAH^{-1}. \tag{26}$$

Computing the action of B_1 requires fewer additions than computing the action of A. To see this, we must describe B_1.

Write

$$\underline{f}(p) = \begin{bmatrix} 1 \\ 0 \end{bmatrix} \otimes 1_{p'/2}. \tag{27}$$

Then

$$(F(2) \otimes I_{p'/2})1_{p'} = 2\underline{f}(p). \tag{28}$$

This leads to the following description of B_1

$$B_1 = \begin{pmatrix} B_1(p) & B_1(p) \otimes \underline{f}^t(q) \\ -2B_1(p) \otimes \underline{f}(q) & B_1(p) \otimes I_{q'} \end{pmatrix}, \tag{29}$$

where $B_1(p)$ is given by

$$B_1(p) = \begin{pmatrix} 1 & \underline{f}^t(p) \\ -2\underline{f}(p) & I_{p'} \end{pmatrix}. \tag{30}$$

We can also compute B_1 by

$$B_1 = Q^{-1}(B_1(q) \otimes B_1(p))Q, \tag{31}$$

where $Q = I_p \otimes P(pq', q')$.

10.8. Summary

Suppose $M = p \cdot q$, p and q distinct primes.

(1). Fundamental Factorization

$$F_\pi = CA$$

where

$$C = 1 \otimes C_p \otimes C_q \otimes (C_p \otimes C_q),$$

$$A = \begin{pmatrix} A(p) & A(p) \otimes 1_{q'}^t \\ A(p) \otimes 1_{q'} & A(p) \otimes I_{q'} \end{pmatrix}, \qquad q' = q - 1.$$

C_p and C_q are rotated Winograd cores determined by the idempotents of Z/N.

(2). Variants

(i)Variant   1

$$F_\pi = FDFA,$$

where D is the diagonal matrix

$$D = 1 \otimes D_p \otimes D_q \otimes D_p \otimes D_q,$$

$$D_p = F(p')^{-1}C_pF(p')^{-1}, \qquad p' = p - 1,$$

and

$$F = 1 \otimes F(p') \otimes F(q') \otimes (F(p') \otimes F(q')).$$

(ii) Variant   2

$$F_\pi = FDBF,$$

where

$$B = FAF^{-1} = \begin{pmatrix} B(p) & B(p) \otimes \underline{e}^t(q) \\ -p'B(p) \otimes \underline{e}(q) & B(p) \otimes I_{q'} \end{pmatrix},$$

and $\underline{e}(q)$ is the q'-tuple

$$\underline{e}(q)^t = \begin{pmatrix} 1 & 0 & \cdots & 0 \end{pmatrix}.$$

Set

$$C_p = \begin{pmatrix} X_p & X_p^* \\ X_p^* & X_p \end{pmatrix},$$

(iii) Variant   3

$$F_\pi = HYHA,$$

where

$$H = 1 \otimes (F(2) \otimes I_{p'/2}) \otimes (F(2) \otimes I_{q'/2}) \otimes ((F(2) \otimes I_{p'/2}) \otimes (F(2) \otimes I_{q'/2})),$$

$$Y = 1 \otimes Y_p \otimes Y_q \otimes (Y_p \otimes Y_q),$$

and

$$Y_p = \frac{1}{2}((X_p + X_p^*) \otimes (X_p - X_p^*)).$$

(iv) Variant   4

$$F_\pi = HYB_1H,$$

where

$$B_1 = HAH^{-1} = \begin{pmatrix} B_1(p) & B_1(p) \otimes \underline{f}^t(q) \\ -2B_1(p) \otimes \underline{f}(q) & B_1(p) \otimes I_{q'} \end{pmatrix},$$

and

$$\underline{f}(q) = \begin{pmatrix} 1 \\ 0 \end{pmatrix} \otimes 1_{q'/2}.$$

Tables of Arithmetic Counts

Table 3.1. $F_\pi = CA$ Direct Method(N=15)

Factor	R.A.	R.M.
A	88	{22}
C	228	272
F_π	316	272+{22}

Table 4.1. $C = HYH$ (N=15)

Factor	R.A.	R.M.
H	44	0
Y	24	68
C	112	68

Table 4.2. $F_\pi = HYB_1H$ (N=15)

Factor	R.A.	R.M.
B_1	44	{11}
F_π	156	68 + {11}

Multiplication by integers has been placed in brackets.

Table 6.1. $F_\pi = CA$ Direct Method(N=pq)

Factor	R.A.	R.M.
A	$4(p'q + pq')$	0
C	$2(p'(2p - 3)q + q'(2q - 3)p)$	$4(p'^2q + q'^2p)$

Table 6.2. $F_\pi = CA$ Direct Method

Size	R.A.	R.M.
15	316	272
21	608	544
35	1284	1168
55	2892	2704

Table 7.1. Real Additions

Size	A	B
15	88	32
21	128	40
35	232	48
55	376	64

Table 7.2. Arithmetic Counts of H and Y

Factor	R.A.	R.M.
H	$2((p-1)q + (q-1)p)$	0
Y	$(p^2-1)q + (q^2-1)p$	$(p-1)^2q + (q-1)^2p$

Computing C by (23) leads to the following table.

Table 7.3. Real Additions for Computing C

Size	Direct Mothod	C=HYH
15	228	112
21	480	200
35	1052	408
55	2516	864

[References]

[1] Good, I. J."The Interaction Algorithm and Practical Fourier Analysis", *J. R. Statist. Soc. B*, vol. 20, No. 2, 1958:pp 361-372.

[2] Thomas, L. H. "Using a Computer to Solve Problems in Physics, Application of Digital Computers", *Ginn and Co.*, Boston, Mass., 1963.

[3] Burrus, C.S. and Eschenbacher, P.W. "An In-place In-order Prime Factor FFT Algorithm", IEEE Trans. on ASSP, Vol. 29, No. 4, pp. 806-817, Aug. 1981.

[4] Chu, S. and Burrus, C.S. "A Prime Factor FFT Algorithm Using Distributed Arithmetic", IEEE Trans. on ASSP, Vol. 30, No. 2, pp. 217-227, April 1982.

[5] Kolba, D.P. and Parks, T.W. " A Prime Factor FFT Algorithm Using High-speed Convolution", IEEE Trans. on ASSP, Vol. 25, No. 4, pp. 281-294, Aug. 1977.

[6] Johnson, R. W. Lu, Chao and Tolimieri, R. "Fast Fourier Transform Algorithms for the Sizes of Product of Distinct Primes and Implementations on VAX", to be published.

[7] Blahut, R. E. *Fast Algorithms for Digital Signal Processing*, Chapter 4 nad 8, Addison-Wesley Pub. Co., 1985.

[8] Nussbaumer, H. J. *Fast Fourier Transform and Convolution Algorithms*, second edition, Chapter 7. Springer-Verlag, 1982.

Problems

1. For $p = 3$ and $q = 7$, find the system of idempotents $\{e_1,\ e_2\}$ which satisfy the conditions given in (1.3) and (1.4).

2. Find the unit group $U(21)$, list all the elements by the ordering defined in Eq. (17) of section 1.

3. Take generator 2 for $U(3)$ and 3 for $U(5)$, order the indexing set $Z/15$, write a complete Fourier transform matrix corresponding to this ordering, and compare with the results in section 2.

4. Prove the tables in section 3, the arithmetic counts of $F_\pi(15)$.

5. Find the system of idempotents for $Z/33$, $Z/35$ and $Z/39$, reorder the indexing set by the found idempotents.

6. Write the explicit matrix of C_3 and C_{11} in $F_\pi(33)$.

7. Write the explicit matrix of C_5 and C_7 in $F_\pi(35)$.

8. Write the explicit matrix of C_3 and C_{13} in $F_\pi(39)$.

9. Prove Eq. (16) and (17) of section 7.

10. Prove the formulas given in Table 2 and 4 of section 7.

Chapter 11

MFTA: TRANSFORM SIZE $N = Mr$

M-COMPOSITE INTEGER AND r-PRIME

11.1. Introduction

In this chapter, we extend the methods introduced in the proceeding two chapters to include the case of transform size N, N a product of three or more distinct primes. In fact, we will give a procedure for designing algorithms for transform size $N = Mr$, M and r relatively prime and r prime, whenever an algorithm for transform size M is given. We will also include FT algorithms for transform size $N = 4M$ where M is a product of distinct odd primes.

Suppose that the transform size N has the form

$$N = pqr, \tag{1}$$

where p, q and r are distinct primes. The set

$$\{e_1, \, e_2, \, e_3\} \tag{2}$$

is a *complete system of idempotents* of Z/N if

$$e_1 \equiv 1 \bmod p, \quad e_1 \equiv 0 \bmod q, e_1 \equiv 0 \bmod r, \tag{3}$$

$$e_2 \equiv 0 \bmod p, \quad e_2 \equiv 1 \bmod q, e_2 \equiv 0 \bmod r, \tag{4}$$

$$e_3 \equiv 0 \bmod p, \quad e_3 \equiv 0 \bmod q, e_3 \equiv 1 \bmod r. \tag{5}$$

Suppose $F_\pi = C_\pi A_\pi$ is the factorization given in the preceeding chapter on data reindexed by the permutation π of Z/M, $M = pq$. The main result of this chapter is the existence of a permutation ρ of Z/N, $N = Mr$, such that

$$F_\rho = C_\rho A_\rho, \tag{6}$$

where

$$A_\rho = \begin{bmatrix} A_\pi & A_\pi \otimes 1^t_{r'} \\ -A_\pi \otimes 1_{r'} & A_\pi \otimes I_{r'} \end{bmatrix}, \quad r' = r - 1, \qquad (7)$$

and

$$C_\rho = 1 \oplus C_p \oplus C_q \oplus (C_p \otimes C_q) \oplus C_r \oplus (C_p \otimes C_r)$$

$$\oplus (C_q \otimes C_r) \oplus (C_p \otimes C_q \otimes C_r). \qquad (8)$$

The matrices C_p, C_q, and C_r are rotated Winograd cores. We will describe these matrices at the end of the section.

Variants of (6) can be designed by the same methods discussed in chapter 10.

It is easy to describe an inductive procedure for extending (6) to transform size N, N a product of more than three distinct primes. Our experience indicates that the more prime factors in the transform size, the more efficient the program as measured by run-time. The main theorem also can be used to design algorithms for transform size $N = 4M$, M a product of distinct odd primes. First we need a 4-point FT algorithm. Although the indexing set $Z/4$ is not a field, the unit group

$$U(4) = \{1, 3\}, \qquad (9)$$

is still cyclic. Order $Z/4$ by the permutation

$$\pi = (0; 2; 1; 3). \qquad (10)$$

Direct computation shows that

$$F_\pi = \begin{bmatrix} 1 & 1 & 1 & 1 \\ 1 & 1 & -1 & -1 \\ 1 & -1 & i & -i \\ 1 & -1 & -i & i \end{bmatrix}, \qquad (11)$$

which we can factorize as

$$F_\pi = C(4)\,A(4), \qquad (12)$$

where

$$C(4) = \begin{bmatrix} 1 & 0 & 0 & 0 \\ 0 & 1 & 0 & 0 \\ 0 & 0 & 1 & i \\ 0 & 0 & 1 & -i \end{bmatrix}, \tag{13}$$

$$A(4) = \begin{bmatrix} 1 & 1 & 1 & 1 \\ 1 & 1 & -1 & -1 \\ 1 & -1 & 0 & 0 \\ 0 & 0 & 1 & -1 \end{bmatrix}. \tag{14}$$

If $N = 4p$, p an odd prime, we will show that a permutation ρ of Z/N can be defined such that

$$F_\rho = C_\rho A_\rho, \tag{15}$$

where

$$A_\rho = \begin{bmatrix} A(4) & A(4) \otimes 1_{p'}^t \\ -A(4) \otimes 1_{p'} & A(4) \otimes I_{p'} \end{bmatrix}, \quad p' = p - 1, \tag{16}$$

$$C_\rho = C'(4) \otimes (C'(4) \otimes C_p). \tag{17}$$

$C'(4)$ is $C(4)$ or $C^*(4)$, the complex conjugate of $C(4)$. C_p is a rotated Winograd core determined by the complete system of idempotents of the ring Z/N.

11.2. Main Theorem

Let $N = Mr$ where M and r are relatively prime and r is prime. Fix a generator z in the unit group $U(r)$ and a permutation π of Z/M. Form the matrix

$$F_\pi = \left[v^{\pi(a)\pi(b)} \right]_{0 \le a, b < M}, \quad v = e^{2\pi i/M}. \tag{1}$$

The action of F_π computes the M-point FT on data reindexed by π. We will obtain an induction procedure for constructing an N-point FT algorithm with F_π embedded in the computation. Consequently, any F_π algorithm can be called upon and used to compute the N-point FT.

By the Chinese Remainder theorem, elements e'_1 and e'_2 can be found in Z/N such that

$$e'_1 \equiv 1 \bmod M, \quad e'_1 \equiv 0 \bmod r, \tag{2}$$

$$e'_2 \equiv 0 \bmod M, \quad e'_2 \equiv 1 \bmod r. \tag{3}$$

Each $a \in Z/N$ can be written uniquely as

$$a \equiv a_1 e'_1 + a_2 e'_2, \quad 0 \leq a_1 < M, \quad 0 \leq a_2 < r. \tag{4}$$

Partition Z/N into two sets

$$S = \{\pi(a_1)e'_1 \mid 0 \leq a_1 < M\}, \tag{5}$$

$$T = \{\pi(a_1)e'_1 + z^k e'_2 \mid 0 \leq a_1 < M, \ 0 \leq k < r - 1\}. \tag{6}$$

We order S by the parameter a_1 and T by the parameters a_1 and k, with k the fastest running parameter. Order Z/N by the permutation

$$\rho = (S, \ T), \tag{7}$$

and consider

$$F_\rho = \left[v_N^{\rho(l)\rho(k)}\right]_{0 \leq l, k < N}. \tag{8}$$

In general, $v_R = exp(2\pi i/R)$. F_ρ computes the N-point FT on the data reindexed by ρ. We decompose F_ρ into the four sub-matrices corresponding to the Cartesian products $S \times S$, $S \times T$, $T \times S$ and $T \times T$. Arguing as in Chapter 10,

$$F_\rho = \begin{bmatrix} F'_\pi & F'_\pi \otimes 1^t_{r'} \\ F'_\pi \otimes 1_{r'} & F'_\pi \otimes C'_r \end{bmatrix}, \quad r' = r - 1, \tag{9}$$

where

$$F'_\pi = \left[(v_N^{e'_1})^{\pi(a_1)\pi(b_1)}\right]_{0 \leq a_1, b_1 < M}, \tag{10}$$

$$C'_r = \left[\left(v_N^{e'_2} \right)^{z^{l+k}} \right]_{0 \le l, k < r-1}. \tag{11}$$

$v_N^{e'_1}$ is a primitive M-th root of unity and $v_N^{e'_2}$ is a primitive r-th root of unity. We form F'_π by replacing v_M in F_π by $v_N^{e'_1}$ and we form C'_r by replacing v_r in $C(r)$ by $v_N^{e'_2}$. Since

$$C'_r 1_{r'} = -1_{r'}, \quad r' = r - 1, \tag{12}$$

we have the *Fundamental Factorization*

$$F_\rho = \left(F'_\pi \ \oplus \ \left(F'_\pi \ \otimes \ C'_r \right) \right) \begin{bmatrix} I_M & I_M \otimes 1^t_{r'} \\ -I_M \otimes 1_{r'} & I_{Mr'} \end{bmatrix}. \tag{13}$$

Formula (13) will be used in the following sections to extend the results of chapter 10. As we see, (13) can be used to embed a factorization of F'_π into a factorization of F_ρ and in this way, gives an inductive procedure for constructing algorithms having any number of distinct prime factors.

11.3. <u>Product of Three Distinct Primes</u>

We will now apply formula (2.13) to the design of FT algorithms for transform size $N = pqr$, where p, q and r are distinct primes. Take $M = pq$ and let π be the permutation of Z/M which leads to the factorization

$$F_\pi = C_\pi A_\pi, \tag{1}$$

given in the preceeding chapter. Set

$$F'_\pi = C'_\pi A_\pi, \tag{2}$$

where C'_π is formed by replacing v_M in C_π by $v_N^{e'_1}$.
Writing

$$F'_\pi \otimes C'_r = \left(C'_\pi \otimes C'_r \right) \left(A_\pi \otimes I_{r'} \right), \quad r' = r - 1, \tag{3}$$

and placing (3) into (2.13), we can write

$$F_\rho = C_\rho \, A_\rho, \tag{4}$$

where

$$A_\rho = (A_\pi \oplus (A_\pi \otimes I_{r'})) \begin{bmatrix} I_M & I_M \otimes 1^t_{r'} \\ -I_M \otimes 1_{r'} & I_{Mr'} \end{bmatrix}, \tag{5}$$

$$C_\rho = C'_\pi \oplus (C'_\pi \otimes C'_r). \tag{6}$$

Direct computation shows that

$$A_\rho = \begin{bmatrix} A_\pi & A_\pi \otimes 1^t_{r'} \\ -A_\pi \otimes 1_{r'} & A_\pi \otimes I_{r'} \end{bmatrix}. \tag{7}$$

Consider C_ρ. Suppose e_1, e_2 and e_3 form a complete system of idempotents of Z/N. Then

$$\{e_1 \bmod M, \ e_2 \bmod M\} \tag{8}$$

is a complete system of idempotents for Z/M. By chapter 10,

$$C_\pi = 1 \oplus C_p \oplus C_q \oplus (C_p \otimes C_q), \tag{9}$$

where C_p is formed by replacing v_p in $C(p)$ by $v_M^{e_1}$. Then, since C'_π is formed by replacing v_M in C_π by $v_N^{e'_1}$, we can write

$$C'_\pi = 1 \oplus C'_p \oplus C'_q \oplus (C'_p \otimes C'_q), \tag{10}$$

where C'_p is formed by replacing v_p in $C(p)$ by $v_N^{e_1 e'_1}$. In the same way, C'_q is formed by replacing v_q in $C(q)$ by $v_N^{e_2 e'_1}$. Placing (10) into (6), we have

$$C_\rho = 1 \oplus C'_p \oplus C'_q \oplus (C'_p \otimes C'_q) \oplus C'_r \oplus (C'_p \otimes C'_r)$$

$$\oplus (C'_q \otimes C'_r) \oplus (C'_p \otimes C'_q \otimes C'_r). \tag{11}$$

From (1.14) and (2.1),

$$e_1 e_1' \equiv 1 \bmod p, \quad e_1 e_1' \equiv 0 \bmod q, \quad e_1 e_1' \equiv 0 \bmod r, \quad (12)$$

implying

$$e_1 \equiv e_1 e_1' \bmod N. \tag{13}$$

It follows that C_p' is formed by replacing v_p in $C(p)$ by $v_N^{e_1}$. Arguing in the same way, C_q' is formed by replacing v_q in $C(q)$ by $v_N^{e_2}$ and C_r' is formed by replacing v_r in $C(r)$ by $v_N^{e_3}$. We see that the rotated Winograd cores C_p', C_q' and C_r' in (11) are completely determined by the complete system of idempotents of Z/N.

Variants of (4) can be produced by the same methods as used in chapter 10. For example, we have

Variant 1.

$$F_\rho = F \, D_\rho \, F \, A_\rho,$$

Variant 2.

$$F_\rho = F \, D_\rho \, B_\rho \, F,$$

where F is the direct sum of small FTs and their tensor products, D_ρ is a diagonal matrix and B_ρ can be iteratively constructed as follows:

$$B(p) = \begin{bmatrix} 1 & \underline{e}^t(p) \\ -p'\underline{e}(p) & I_{p'} \end{bmatrix}, \tag{14}$$

$$B_\pi = \begin{bmatrix} B(p) & B(p) \otimes \underline{e}^t(q) \\ -q'B(p) \otimes \underline{e}(q) & B(p) \otimes I_{q'} \end{bmatrix}, \tag{15}$$

$$B_\rho = \begin{bmatrix} B_\pi & B_\pi \otimes \underline{e}^t(r) \\ -r'B_\pi \otimes \underline{e}(r) & B_\pi \otimes I_{r'} \end{bmatrix}. \tag{16}$$

11.4. Transform Size $N = 12$

We will apply (2.13) with $M = 4$, $r = 3$ and

$$\pi = (0,\ 2,\ 1,\ 3). \tag{1}$$

In the introduction, we determined the factorization

$$F_\pi = C(4)A(4). \tag{2}$$

The idempotents in this case are

$$e'_1 = 9, \quad e'_2 = 4. \tag{3}$$

Since $v^9_{12} = -i$, the matrix F'_π in (15) of section 2. is

$$F'_\pi = C^*(4)A(4), \tag{4}$$

The permutation ρ in (2.13) is

$$\rho = (0,\ 6,\ 9,\ 3;\quad 4,\ 8,\ 10,\ 2;\quad 1,\ 5,\ 7,\ 11), \tag{5}$$

and we have by (15) of section 2,

$$F_\rho = (F'_\pi \oplus (F'_\pi \otimes C_3)) \begin{bmatrix} I_4 & I_4 \otimes 1^t_2 \\ -I_4 \otimes 1_2 & I_8 \end{bmatrix}. \tag{6}$$

Since $v^4_{12} = v_3$, $C_3 = C(3)$. Arguing as in the preceeding section

$$F_\rho = C_\rho A_\rho, \tag{7}$$

where

$$A_\rho = \begin{bmatrix} A(4) & A(4) \otimes 1^t_2 \\ -A(4) \otimes 1_2 & A(4) \otimes I_2 \end{bmatrix}, \tag{8}$$

$$C_\rho = C(4)^* \oplus (C(4)^* \otimes C(3)). \tag{9}$$

Arithmetically A_ρ is equivalent to

$$A(4) \otimes A(3). \tag{10}$$

11.5. <u>Transform Size</u> $N = 4p$

Choose a complete system of idempotents for the ring Z/N, $N = 4p$, p an odd prime,

$$\{e_1', \, e_2'\}. \tag{1}$$

Again, we take

$$\pi = (0, \, 2; \, 1, \, 3), \tag{2}$$

in (2.15)

$$F_\pi' = C(4) \, A(4), \quad v_N^{e_1'} = i, \tag{3}$$

or

$$F_\pi' = C^*(4) \, A(4), \quad v_m^{e_1'} = -i. \tag{4}$$

The permutation ρ in (2.15) is given by

$$0, \, 2e_1, \, e_1, \, 3e_1, \tag{5}$$

$$e_2, \, ze_2, \, \cdots, \, z^{p-2}e_2, \tag{6}$$

$$2e_1 + e_2, 2e_1 + ze_2, \cdots, \, 2e_1 + z^{p-2}e_2, \tag{7}$$

$$e_1 + e_2, \, e_1 + ze_2, \cdots, \, e_1 + z^{p-2}e_2, \tag{8}$$

$$3e_1 + e_2, \, 3e_1 + ze_2, \cdots, \, 3e_1 + z^{p-2}e_2, \tag{9}$$

where z is a generator of $U(p)$. Applying (2.13),

$$F_\rho = (F_\pi' \oplus (F_\pi' \otimes C_p)) \begin{bmatrix} I_4 & I_4 \otimes 1_{p'}^t \\ -I_4 \otimes 1_{p'} & I_{4p'} \end{bmatrix}, \, p' = p - 1, \tag{10}$$

which by direct computation using (3) or (4) can be rewritten as

$$F_\rho = C_\rho \, A_\rho, \tag{11}$$

where

$$A_\rho = \begin{bmatrix} A(4) & A(4) \otimes 1_{p'}^t \\ -A(4) \otimes 1_{p'} & A(4) \otimes I_{p'} \end{bmatrix}, \qquad (12)$$

$$C_\rho = C_4 \oplus (C_4 \otimes C_p). \qquad (13)$$

$C_4 = C(4)$ or $C^*(4)$ depending on whether $v_N^{e_1'} = i$ or $v_N^{e_1'} = -i$, and C_p is formed by replacing v_p in $C(p)$ by $v_N^{e_2'}$.

Arithmetically, A_ρ is equivalent to

$$A(4) \otimes A(p). \qquad (14)$$

11.6. Variants $N = 4p$

Several variants, in the spirit of chapter 10, will now be designed. Set

$$F = I_4 \oplus (I_4 \otimes F(p')), \quad p' = p - 1, \qquad (1)$$

and define

$$D_\rho = F^{-1} C_\rho F^{-1}, \qquad (2)$$

and

$$D_p = F(p')^{-1} C_p F(p')^{-1}. \qquad (3)$$

Placing D_ρ in (5.11), we have our first variant

Variant 1.

$$F_\rho = F D_\rho F A_\rho. \qquad (4)$$

To reduce the number of additions required to implement (5.11) and Variant 1, we rewrite Variant 1 as

Variant 2.

$$F_\rho = F D_\rho B_\rho F, \qquad (5)$$

where by direct computation

$$B_\rho = F A_\rho F^{-1} = \begin{bmatrix} A(4) & A(4) \otimes \underline{e}^t(p) \\ -p' A(4) \otimes \underline{e}(p) & A(4) \otimes I_{p'} \end{bmatrix}, \qquad (6)$$

where $\underline{e}(p)$ is the p'-tuple given by

$$\underline{e}^t(p) = \begin{bmatrix} 1 & 0 & \cdots & 0 \end{bmatrix}. \tag{7}$$

Arithmetically, B_ρ is equivalent to

$$A(4) \otimes B(p), \tag{8}$$

where

$$B(p) = \begin{bmatrix} 1 & \underline{e}^t(p) \\ -p'\underline{e}(p) & I_{p'} \end{bmatrix}. \tag{9}$$

The next variant replaces the complex multiplications required to compute the action of C_ρ by real multiplications and produces the most balanced algorithm as measured by the required additions and multiplications. First observe C_p has the form

$$C_p = \begin{bmatrix} X_p & X_p^* \\ X_p^* & X_p \end{bmatrix}. \tag{10}$$

with the result that

$$C_p = (F(2) \otimes I_{p'/2}) \, Y_p \, (F(2) \otimes I_{p'/2}), \tag{11}$$

where

$$Y_p = 1/2((X_p + X_p^*) \oplus (X_p - X_p^*)). \tag{12}$$

The entries of Y_p are all real in the block $X_p + X_p^*$ and all purely imaginary in the block $X_p - X_p^*$. We can now write

$$C_\rho = H \, Y_\rho \, H, \tag{13}$$

where

$$H = I_4 \oplus (I_4 \otimes F(2) \otimes I_{p'/2}), \tag{14}$$

$$Y_\rho = C_4 \oplus (C_4 \otimes Y_p). \tag{15}$$

Placing (14) into the factorization (7) of section 5., we have the variant

Variant 3.

$$F_\rho = H \, Y_\rho \, H \, A_\rho. \tag{16}$$

Again we can reduce the additions by the variant;

Variant 4.

$$F_\rho = H \, Y_\rho \, B'_\rho \, H, \tag{17}$$

where

$$B'_\rho = H \, A_\rho \, H^{-1} = \begin{bmatrix} A(4) & A(4) \otimes \underline{f}^t(p) \\ -2A(4) \otimes \underline{f}(p) & A(4) \otimes I_{p'} \end{bmatrix}, \tag{18}$$

$$\underline{f}(p) = \begin{bmatrix} 1 \\ 0 \end{bmatrix} \otimes 1_{\frac{p'}{2}}. \tag{19}$$

Arithmetically B'_ρ is equivalent to

$$A(4) \otimes B_1(p), \tag{20}$$

where $B_1(p)$ is defined in chapter 9 by

$$B_1(p) = \begin{bmatrix} 1 & \underline{f}^t(p) \\ -2\underline{f}(p) & I_{p'} \end{bmatrix}. \tag{21}$$

The number of multiplications by small integers are placed in the brackets { }.

11.7. <u>Transform Size $N = 60$</u>

We apply (2.13) to the case $N = 60$ where $M = 12$ and $r = 5$. Let π in (2.13) be the permutation ρ defined by (4.5). Then, by the results of section 5.,

$$F_\pi = C_\pi \, A_\pi, \tag{1}$$

where

$$A_\pi = \begin{bmatrix} A(4) & A(4) \otimes 1_2^t \\ -A(4) \otimes 1_2 & A(4) \otimes I_2 \end{bmatrix}. \tag{2}$$

Arguing as in section 3., with ρ defined by (2.7),

$$F_\rho = C_\rho A_\rho, \tag{4}$$

where

$$A_\rho = \begin{bmatrix} A_\pi & A_\pi \otimes 1_4^t \\ -A_\pi \otimes 1_4 & A_\pi \otimes I_4 \end{bmatrix}, \tag{5}$$

$$C_\rho = C'_\pi \oplus (C'_\pi \otimes C_5). \tag{6}$$

The rotated Winograd cores making up

$$C'_\pi = C_4 \oplus (C_4 \otimes C_3), \tag{7}$$

and the Winograd core C_5 are completely determined by the complete system of idempotents

$$e_1 = 45, \quad e_2 = 40, \quad e_3 = 36, \tag{8}$$

of the ring $Z/60$. Since

$$v_{60}^{45} = -i, \quad v_{60}^{40} = v_3^2, \quad v_{60}^{36} = v_5^3, \tag{9}$$

we have $C_4 = C^*(4)$, C_3 is formed by replacing v_3 in $C(3)$ by v_3^2 and C_5 is formed by replacing v_5 in $C(5)$ by v_5^3.

Tables of Arithmetic Counts

Table 4.1. $F_\rho = C_\rho A_\rho$

Factor	R.A.	R.M.
C_ρ	60	64
A_ρ	68	0
F_ρ	128	64

Table 5.1. $F_\rho = C_\rho A_\rho$

Factor	R.A.	R.M.
C_ρ	4(p-1)(4p-5)+4	$16(p-1)^2$
A_ρ	4(7p-4)	0
F_ρ	$4(4p^2 - 2p + 2)$	$16(p-1)^2$

Table 5.2. $F_\rho = C_\rho A_\rho$

Factor	R.A.	R.M.
12	128	64
20	368	256
28	736	576
44	1856	1600

Table 6.1. C=HYH (N=4p)

Factor	R.A.	R.M.
Y	$4(p^2 - 4p + 6)$	$4(p-1)^2$
H	8(p-1)	0
C	$4(p^2 + 2)$	$4(p-1)^2$

Table 6.2. $F_\pi = HY B_1 H$

Factor	R.A.	R.M.
B_1	20p-8	$\{4(p-1)\}$
F_π	4p(p+5)	$4(p-1)^2 + \{4(p-1)\}$

Table 6.3. $F_\pi = HY B_1 H$

Factor	R.A.	R.M.
12	96	16+{8}
20	200	64+{16}
28	336	144+{24}
44	704	400+{40}

R.A. – the number of real additions.

R.M. – the number of real multiplications.

[References]

[1] Blahut, R. E. *Fast Algorithms for Digital Signal Processing*, Chapter 6, 8. Addison-Wesley, 1985.

[2] Johnson, R.W. Lu, Chao and Tolimieri, R. "Fast Fourier Algorithms for the Size of 4p and 4pq and Implementations on VAX", submitted for publication.

[3] Lu, Chao *Fast Fourier Transform Algorithms for Special N's and the Implementations on VAX*, Ph.D. Dissertation. Jan., 1988, the City University of New York.

[4] Nussbaumer, H. J. *Fast Fourier Transform and Convolution Algorithms*, second edition, Chapter 7, Springer-Verlag,1982.

Problems

1. For $p = 3$, $q = 5$ and $r = 7$, find the system of idempotents $\{e_1, e_2, e_3\}$ which satisfy (14), (15) and (16) of section 1.

2. Find the ordering of $Z/105$ by the idempotents of Problem 1.

3. Define the matrices C_3, C_5 and C_7 in $F(105)$.

4. Derive the $N = 4p$ algorithm for $p = 5$, write in detail as the example of $N = 12$ as given in section 4.

5. Derive the four variants of the $N = 20$ algorithm.

6. For prime p with what property such that the F'_π has to be written as $F'_\pi = C^*(4)A(4)$.

7. Prove the formulas given in table 1 of section 5.

8. Prove the formulas in table 1 and 2 of section 6.

Chapter 12

MFTA: TRANSFORM SIZE $N = p^2$

12.1. Introduction

In chapters 8 to 11, multiplicative algorithms were designed for the FFT of transform sizes N, N a prime, a product of distinct primes and a product of relatively primes. These algorithms start with the multiplicative ring-structure of the indexing set, in the spirit of the Good-Thomas PFA and compute the resulting factorization by combining Rader and Winograd small FFT algorithms. The basic factorization is

$$F_\pi = CA, \tag{1}$$

where C is a block diagonal matrix with small skew-circulant blocks (rotated Winograd cores) and tensor product of these small skew-circulant blocks , and A is a pre-addition matrix with all its entries being 0, 1 or -1.

In this chapter, we take up the case of transform size N, $N = p^2$ and $N = p^k$, p an odd prime. The indexing of input and output data are the same as the Winograd algorithm, but the factorization follows the structure of (1) after a modification. Two different algorithms are given for the case of $N = p^2$ and examples are presented in detail. In section 2, we start with an example of $N = p^2$, $N = 9$. The general case $N = p^2$ will be given in detail in section 3. An extension to the case of p^k is given in section 4 by the example of $N = 3^3 = 27$.

12.2. An Example N=9

Algorithms for computing 9-point Fourier transform will be designed from the ring-structure of the indexing set $Z/9$. Denote the unit group of $Z/9$ by U, then

$$U = \{1,\ 2,\ 4,\ 5,\ 7,\ 8\}. \tag{1}$$

Reordering U by a generator $z = 2$, we have

$$D_0 = U = \{1,\ 2,\ 4,\ 8,\ 7,\ 5\}. \tag{2}$$

The unit group operating on $Z/9$ has orbits D_0, D_1 and D_2, where

$$D_1 = 3U = \{3,\ 6\}, \tag{3}$$

$$D_2 = 0U = \{0\}. \tag{4}$$

This provides a reindexing of $Z/9$ defined by the permutation

$$\pi = (0;\ \ 3,6;\ \ 1,2,4,8,7,5). \tag{5}$$

Denote the corresponding permutation matrix by P. Then,

$$F_\pi(9) = PF(9)P^{-1}, \tag{6}$$

where

$$F_\pi(9) = \begin{bmatrix} 1 & 1 & 1 & 1 & 1 & 1 & 1 & 1 & 1 \\ 1 & 1 & 1 & v & v^2 & v & v^2 & v & v^2 \\ 1 & 1 & 1 & v^2 & v & v^2 & v & v^2 & v \\ 1 & v & v^2 & w & w^2 & w^4 & w^8 & w^7 & w^5 \\ 1 & v^2 & v & w^2 & w^4 & w^8 & w^7 & w^5 & w \\ 1 & v & v^2 & w^4 & w^8 & w^7 & w^5 & w & w^2 \\ 1 & v^2 & v & w^8 & w^7 & w^5 & w & w^2 & w^4 \\ 1 & v & v^2 & w^7 & w^5 & w & w^2 & w^4 & w^8 \\ 1 & v^2 & v & w^5 & w & w^2 & w^4 & w^8 & w^7 \end{bmatrix},$$

$$w = exp(2\pi i/9), \quad and \quad v = exp(2\pi i/3). \tag{7}$$

We rewrite (7) as

$$F_\pi(9) = \begin{bmatrix} 1 & 1_2^t & 1_6^t \\ 1_2 & E(2,2) & 1_3^t \otimes C_3 \\ 1_6 & 1_3 \otimes C_3 & C_9 \end{bmatrix}. \tag{8}$$

where

$$C_3 = \begin{pmatrix} v & v^2 \\ v^2 & v \end{pmatrix}, \tag{9}$$

$$C_9 = \begin{pmatrix} w & w^2 & w^4 & w^8 & w^7 & w^5 \\ w^2 & w^4 & w^8 & w^7 & w^5 & w \\ w^4 & w^8 & w^7 & w^5 & w & w^2 \\ w^8 & w^7 & w^5 & w & w^2 & w^4 \\ w^7 & w^5 & w & w^2 & w^4 & w^8 \\ w^5 & w & w^2 & w^4 & w^8 & w^7 \end{pmatrix}, \tag{10}$$

and $E(n,n)$ is $n \times n$ matrix with all of whose entries are 1's, 1_n is n dimensional vector having all components equal to 1, and superscript t denotes the transpose matrix. By O_n we mean the n-dimensional vector having all components equal to 0. Since $C_9 1_6 = O_6$, we cannot directly factor out C_9 from $F_\pi(9)$. The following two algorithms I and II present methods for getting around this problem.

2.1 Algorithm I $N = 9$

Let $0(n,n)$ denote the $n \times n$ matrix with all 0's, and use $+$ to denote matrix addition. Modify $F_\pi(9)$ by adding to it the matrix $0(3,3) \oplus (\frac{1}{3}E(3,3) \otimes C_3)$ to form

$$F_c(9) = F_\pi(9) + 0(3,3) \oplus \left(\frac{1}{3}E(3,3) \otimes C_3\right). \tag{11}$$

Then $F_c(9)$ is given as

$$F_c(9) = \begin{bmatrix} 1 & 1_2^t & 1_6^t \\ 1_2 & E(2,2) & 1_3^t \otimes C_3 \\ 1_6 & 1_3 \otimes C_3 & C_9' \end{bmatrix}, \tag{12}$$

where

$$C_9' = C_9 + \frac{1}{3}E(3,3) \otimes C_3. \tag{13}$$

The matric C'_9 is skew-circulant, and

$$C'_9 1_6 = -1_6. \tag{14}$$

The method of the preceeding chapters can be applied to $F_c(9)$. We require the following identities.

$$(E(3,3) \otimes C_3)1_6 = -3\,1_6, \tag{15}$$

$$C_9(1_3 \otimes I_2) = 0(6,6), \tag{16}$$

$$(E(3,3) \otimes C_3)(1_3 \otimes I_2) = 3(1_3 \otimes C_3). \tag{17}$$

$F_c(9)$ can be factorized as

$$F_c(9) = CA, \tag{18}$$

where C is a block-diagonal matrix

$$C = 1 \oplus C_3 \oplus C'_9, \tag{19}$$

and A is the pre-addition matrix

$$A = \begin{bmatrix} 1 & 1_2^t & 1_6^t \\ -1_2 & -E(2,2) & 1_3^t \otimes I_2 \\ -1_6 & 1_3 \otimes I_2 & I_6 \end{bmatrix}. \tag{20}$$

We can write

$$F_\pi(9)\underline{a} = F_c(9)\underline{a} - \begin{bmatrix} 0_3 \\ 1_3 \otimes \underline{b}' \end{bmatrix}, \tag{21}$$

where

$$\underline{b}' = \frac{1}{3}C_3 \begin{bmatrix} a_3 + a_5 + a_7 \\ a_4 + a_6 + a_8 \end{bmatrix}.$$

Variants of the factorization $F_c(p) = CA$ can be obtained by methods in the preceeding chapters.

2.2 Algorithm II $N = 9$

Set

$$H = 1 \oplus F(2) \oplus (F(2) \otimes I_3),$$ (22)

and define $F'_\pi(9)$ by

$$F_\pi(9) = H F'_\pi(9) H.$$ (23)

Direct computation from (7) shows that $F'_\pi(9)$ is the following matrix.

$$
\begin{bmatrix}
1 & 1 & 0 & 1 & 1 & 1 & 0 & 0 & 0 \\
1 & 1 & 0 & -\frac{1}{2} & -\frac{1}{2} & -\frac{1}{2} & 0 & 0 & 0 \\
0 & 0 & 0 & 0 & 0 & 0 & \frac{\alpha}{2} & -\frac{\alpha}{2} & \frac{\alpha}{2} \\
1 & -\frac{1}{2} & 0 & & & & & & \\
1 & -\frac{1}{2} & 0 & & \frac{X_6+X_6^*}{2} & & & 0 & \\
1 & -\frac{1}{2} & 0 & & & & & & \\
1 & 0 & \frac{\alpha}{2} & & & & & & \\
1 & 0 & -\frac{\alpha}{2} & & 0 & & & \frac{X_6-X_6^*}{2} & \\
1 & 0 & \frac{\alpha}{2} & & & & & &
\end{bmatrix},
$$ (24)

where $\alpha = v - v^*$ with $v = exp(2\pi i/3)$ and

$$
X_6 = \begin{bmatrix}
w & w^2 & w^4 \\
w^2 & w^4 & w^8 \\
w^4 & w^8 & w^7
\end{bmatrix}, \quad w = exp(2\pi i/9).
$$ (25)

We will now see how to factor $F'_\pi(9)$. Set

$$T = 1 \oplus 1 \oplus \alpha \oplus I_3 \oplus \alpha I_3,$$ (26)

then

$$F'_\pi(9) = T F''_\pi(9),$$ (27)

where F''_π is the matrix

$$
\begin{bmatrix}
1 & 1 & 0 & 1 & 1 & 1 & 0 & 0 & 0 \\
1 & 1 & 0 & -\frac{1}{2} & -\frac{1}{2} & -\frac{1}{2} & 0 & 0 & 0 \\
0 & 0 & 0 & 0 & 0 & 0 & \frac{1}{2} & -\frac{1}{2} & \frac{1}{2} \\
1 & -\frac{1}{2} & 0 & & & & & & \\
1 & -\frac{1}{2} & 0 & & \frac{X_6 + X_6^*}{2} & & & 0 & \\
1 & -\frac{1}{2} & 0 & & & & & & \\
1 & 0 & \frac{1}{2} & & & & & & \\
1 & 0 & -\frac{1}{2} & & 0 & & & \alpha^{-1}\frac{X_6 - X_6^*}{2} & \\
1 & 0 & \frac{1}{2} & & & & & &
\end{bmatrix} . \tag{28}
$$

We see that $F''_\pi(9)$ is a pure real matrix. Putting all this together, we have

$$F_\pi(9) = H T F''_\pi(9) H. \tag{29}$$

A second derivation of factorization (29) will now be given, which will be generalized in the next section. A direct computation shows that

$$(C_3 \otimes I_3)(1_3 \otimes I_2) = 1_3 \otimes C_3. \tag{30}$$

We factor $F_\pi(9)$ as

$$F_\pi(9) = CA, \tag{31}$$

where

$$C = 1 \oplus C_3 \oplus (C_3 \otimes I_3), \tag{32}$$

and

$$
A = \begin{pmatrix}
1 & 1_2^t & 1_6^t \\
-1_2 & -E(2,2) & 1_3^t \otimes l_2 \\
-1_6 & 1_3 \otimes I_2 & Z
\end{pmatrix}, \tag{33}
$$

with

$$Z = (C_3^{-1} \otimes I_3)C_9. \tag{34}$$

We will now compute Z. First

$$C_3^{-1} = \frac{1}{\alpha} \begin{pmatrix} v & -v^2 \\ -v^2 & v \end{pmatrix}, \quad \alpha = v - v^2. \tag{35}$$

Direct computation shows that

$$Z = \frac{1}{\alpha} \begin{pmatrix} vX_6 - v^2 X_6^* & vX_6^* - v^2 X_6 \\ vX_6^* - v^2 X_6 & vX_6 - v^2 X_6^* \end{pmatrix}. \tag{36}$$

We see that Z is a real matrix. In fact, Z is a 6×6 skew-circulant matrix having first row as

$$\frac{1}{\alpha}[w^4 - w^5, \; w^5 - w^4, \; w^7 - w^2, \; w^2 - w^7, \; w - w^8, \; w^8 - w]. \tag{37}$$

From (31) we can write

$$F_\pi(9) = H(H^{-1}CH)(H^{-1}AH^{-1})H,$$

where $H^{-1}CH$ is the matrix T given in (26) and $H^{-1}AH^{-1}$ is the matrix F_π'' given in (28).

12.3. The General Case $N = p^2$

For the case of $N = p^2$, p an odd prime, Z/N is no longer a field. However, the unit group

$$U = U(p^2) = \{0 \le l < p^2 \mid (l, p^2) = 1\}, \tag{1}$$

is still a cyclic group. Define the subsets D_k of Z/p^2, $0 \le k < 3$ by

$$D_k = \{0 \le l < p^2 \mid \quad (l, p^2) = p^k\}. \tag{2}$$

We have

$$D_0 = U, \tag{3}$$

$$D_1 = \{p, \; 2p, \; \ldots, \; (p-1)p\}, \tag{4}$$

$$D_2 = \{0\}. \tag{5}$$

It is easy to see that D_1 has $(p-1)$ elements and D_2 has 1 element, so that D_0 has an order $p(p-1)$. Since U is a cyclic group, we can find a generator, $z \in U$, such that

$$U = \{z^l \mid \quad 0 \leq l < p(p-1)\}. \tag{6}$$

By Fermat's theorem [1.3], we see that

$$z^{p-1} \equiv 1 \bmod p. \tag{7}$$

Multiplying both sides of (7) by p, we have

$$pz^{p-1} \equiv p \bmod p^2. \tag{8}$$

The elements of D_1 are all of the form pz^l. Since the order of D_1 is $(p-1)$, each element $x \in D_1$ has a unique representation

$$x = pz^l : \quad 0 \leq l < p-1, \tag{9}$$

or

$$D_1 = \{pz^l \mid \quad 0 \leq l < p-1\}. \tag{10}$$

Equations (6) and (10) define a partition of Z/p^2. Ordering the partition by the set of U-orbits

$$D_2, \ D_1, \ D_0, \tag{11}$$

$F(p^2)$ can be written as

$$F(p^2) = P^{-1}F_\pi(p^2)P, \tag{12}$$

where P is the permutation matrix corresponding to permutation π defined in (11), and

$$F_\pi(p^2) = [\, w^{\pi(l)\pi(k)} \,]_{0 \leq l,k < p^2}, \quad w = e^{2\pi i/p^2}. \tag{13}$$

Let $p' = p - 1$, and $s = p(p-1)$. $F_\pi(p^2)$ can be written as

$$F_\pi(p^2) = \begin{bmatrix} 1 & 1_{p'}^t & 1_s^t \\ 1_{p'} & E(p',p') & 1_p^t \otimes C_p \\ 1_s & 1_p \otimes C_p & C_{p^2} \end{bmatrix}, \tag{14}$$

where C_{p^2} is the $s \times s$ skew-circulant matrix having the first row

$$w^{z^0}, w^{z^1}, \ldots, w^{z^{s-1}}, \tag{15}$$

and C_p is the $(p-1) \times (p-1)$ skew-circulant matrix having first row

$$w^{pz^0}, w^{pz^1}, \ldots, w^{pz^{p-2}}, \tag{16}$$

$E(p',p')$ is the p' by p' matrix with all elements equal 1.

3.1 Algorithm I $N = p^2$

The matrix

$$F_c(p^2) = F_\pi(p^2) + 0(p,p) \oplus \left(\frac{1}{p}E(p,p) \otimes C_p\right) \tag{17}$$

can be written in the form

$$F_c(p^2) = \begin{bmatrix} 1 & 1_{p'}^t & 1_s^t \\ 1_{p'} & E(p',p') & 1_p^t \otimes C_p \\ 1_s & 1_p \otimes C_p & C_{p^2} + \frac{1}{p}E(p,p) \otimes C_p \end{bmatrix}. \tag{18}$$

Let

$$C_{p^2}' = C_{p^2} + \frac{1}{p}E(p,p) \otimes C_p. \tag{19}$$

Direct computation shows that

$$C_{p^2} 1_s = 0_s, \quad C_{p^2}' 1_s = 1_s, \tag{20}$$

$$(E(p,p) \otimes C_p)1_s = -p1_s, \tag{21}$$

$$C_{p^2}\left(1_p \otimes I_{p-1}\right) = 0(s,s), \tag{22}$$

$$(E(p,p) \otimes C_p)(1_p \otimes I_{p-1}) = p(1_p \otimes C_p). \qquad (23)$$

Fundamental Factorization I

From these identities, we have the factorization

$$F_c(p^2) = CA, \qquad (24)$$

where C is a block-diagonal matrix

$$C = 1 \oplus C_p \oplus C'_{p^2}, \qquad (25)$$

and A is the pre-addition matrix given as

$$A = \begin{bmatrix} 1 & 1^t_{p'} & 1^t_s \\ -1_{p'} & -E(p',p') & 1^t_p \otimes I_{p'} \\ -1_s & 1_p \otimes I_{p'} & I_s \end{bmatrix}. \qquad (26)$$

We can use (17) to compute the action of $F_\pi(p^2)$.
The arithmetic count is given in the following table.

Variants of the fundamental factorization are found by the usual methods. Details can be found in [5].

3.2 Algorithm II $N = p^2$

We can write

$$C_p = \begin{bmatrix} X_p & X^*_p \\ X^*_p & X_p \end{bmatrix}, \qquad (27)$$

from which it follows that

$$C_p = (F(2) \otimes I_{p'/2}) Y_p (F(2) \otimes I_{p'/2}), \qquad (28)$$

where

$$Y_p = \frac{1}{2} \begin{bmatrix} X_p + X^*_p & 0 \\ 0 & X_p - X^*_P \end{bmatrix}. \qquad (29)$$

Since the matrix C_p is non-singular, it follows that $X_p + X^*_p$ and $X_p - X^*_p$ are non-singular. Direct computation shows that

$$(C_p \otimes I_p)(1_p \otimes I_{p'}) = 1_p \otimes C_p, \qquad (30)$$

which leads to the factorization.

Fundamental Factorization II

$$F_\pi(p^2) = CA, \tag{31}$$

where

$$C = 1 \oplus C_p \oplus (C_p \otimes I_p), \tag{32}$$

and

$$A = \begin{pmatrix} 1 & 1_{p'}^t & 1_s^t \\ -1_{p'} & -E(p',p') & 1_p^t \otimes I_{p'} \\ -1_s & 1_p \otimes I_{p'} & Z \end{pmatrix}, \tag{33}$$

with

$$Z = (C_p^{-1} \otimes I_p)C_{p^2}. \tag{34}$$

Set

$$L = X_p^2 - X_p^{*2}, \tag{35}$$

L is a non-singular matrix and we have

$$C_p^{-1} = (I_2 \otimes L^{-1}) \begin{pmatrix} X_p & -X_p^* \\ -X_p^* & X_p \end{pmatrix}. \tag{36}$$

Set $L_1 = L \otimes I_p$. From (34) and (36), we have

$$Z = (I_2 \otimes L_1^{-1}) \begin{pmatrix} Z_1 & Z_2 \\ Z_2 & Z_1 \end{pmatrix}, \tag{37}$$

where

$$Z_1 = (X_p \otimes I_p)X_s - (X_p^* \otimes I_p)X_s^*, \tag{38}$$

and

$$Z_2 = (X_p \otimes I_p)X_s^* - (X_p^* \otimes I_p)X_s. \tag{39}$$

We see that Z is a real matrix.

Set

$$H = 1 \oplus \left(F(2) \otimes I_{\frac{p'}{2}}\right) \oplus \left(F(2) \otimes I_{\frac{s}{2}}\right). \tag{40}$$

Direct computation shows that

$$H^{-1}AH^{-1} = B, \tag{41}$$

is the matrix

$$
\begin{bmatrix}
1 & 1^t_{\frac{p'}{2}} & 0^t_{\frac{p'}{2}} & 1^t_{\frac{q}{2}} & 0^t_{\frac{q}{2}} \\
-1_{\frac{p'}{2}} & -E(\frac{p'}{2},\frac{p'}{2}) & 0(\frac{p'}{2},\frac{p'}{2}) & & \\
& & & \frac{1}{2}M^t & \\
0_{\frac{p'}{2}} & 0(\frac{p'}{2},\frac{p'}{2}) & 0(\frac{p'}{2},\frac{p'}{2}) & & \\
-1_{\frac{q}{2}} & & & \frac{1}{2}L_1^{-1}(Z_1 + Z_2) & 0 \\
& \frac{1}{2}M & & & \\
0_{\frac{q}{2}} & & & 0 & \frac{1}{2}L_1^{-1}(Z_1 - Z_2)
\end{bmatrix}
\tag{42}
$$

$$
M^t = \begin{pmatrix}
I_{\frac{p'}{2}} & \cdots & I_{\frac{p'}{2}} & 0 & \cdots & 0 & 0 \\
0 & \cdots & 0 & I_{\frac{p'}{2}} & -I_{\frac{p'}{2}} & \cdots & I_{\frac{p'}{2}}
\end{pmatrix}. \tag{43}
$$

The matrix B is a real matrix. In fact

$$L_1^{-1}(Z_1 + Z_2) = ((X_p + X_p^*)^{-1} \otimes I_p)(X_{p^2} + X_{p^2}^*), \tag{44}$$

$$L_1^{-1}(Z_1 - Z_2) = ((X_p - X_p^*)^{-1} \otimes I_p)(X_{p^2} - X_{p^2}^*). \tag{45}$$

We can now write

$$\textbf{Variant 1.} \quad F_\pi(p^2) = CHBH, \tag{46}$$

and

$$\textbf{Variant 2.} \quad F_\pi(p^2) = HDBH, \tag{47}$$

where

$$D = H^{-1}CH, \tag{48}$$

is the block-diagonal matrix

$$D = 1 \oplus (X_p + X_p^*) \oplus (X_p - X_p^*) \oplus ((X_p + X_p^*) \otimes I_p) \oplus ((X_p - X_p^*) \otimes I_p). \tag{49}$$

The blocks of D are either real or purely imaginary.

The arithmetic of variant 2 is given in the following table.

12.4. __An Example N=3^3__

Take $z = 2$ as a generator of the unit group U of $Z/27$,

$$U = \{2^k : \quad 0 \leq k < 18\}. \tag{1}$$

The indexing set $Z/27$ is partitioned by the U-orbits

$$D_0 = 0U = \{0\}, \tag{2}$$

$$D_1 = 9U = \{9,\ 18\}, \tag{3}$$

$$D_2 = 3U = \{3,\ 6,\ 12,\ 24,\ 21,\ 15\}, \tag{4}$$

$$D_3 = U. \tag{5}$$

Consider the permutation π of $Z/27$ defined by

$$\pi = \{D_0;\ D_1;\ D_2;\ D_3\}. \tag{6}$$

The matrix

$$F_\pi = PF(27)P^{-1}, \tag{7}$$

is given by

$$F_\pi = \begin{bmatrix} 1 & 1_2^t & 1_6^t & 1_{18}^t \\ 1_2 & E(2,2) & E(2,6) & 1_9^t \otimes C_3 \\ 1_6 & E(6,2) & E(3,3) \otimes C_3 & 1_3^t \otimes C_9 \\ 1_{18} & 1_9 \otimes C_3 & 1_3 \otimes C_9 & C_{27} \end{bmatrix}, \tag{8}$$

where P is the permutation matrix corresponding to π, C_3 and C_9 are as defined in section 2, and C_{27} is

$$C_{27} = [w^{z^{l+k}}], \qquad w = e^{2\pi i/27}, \tag{9}$$

Set

$$C = 1 \oplus C_3 \oplus (C_3 \otimes I_3) \oplus (C_3 \otimes I_9). \tag{10}$$

The factorization of C from F_π is governed by the following formulas which come from direct computation.

$$(C_3 \otimes I_3)(1_3 \otimes I_2) = 1_3 \otimes C_3, \tag{11}$$

$$(C_3 \otimes I_9)(1_9 \otimes I_2) = 1_9 \otimes C_3, \tag{12}$$

$$Z(9) = (C_3^{-1} \otimes I_3)C_9, \tag{13}$$

$$1_3 \otimes Z(9) = (C_3^{-1} \otimes I_9)(1_3 \otimes C_9), \tag{14}$$

where $Z(9)$ is the matrix Z in section 2. Then

$$F_\pi = CA, \tag{15}$$

where

$$A = \begin{bmatrix} 1 & 1_2^t & 1_6^t & 1_{18}^t \\ -1_2 & -E(2,2) & -E(2,6) & 1_9^t \otimes I_2 \\ -1_6 & -E(6,2) & E(3,3) \otimes I_2 & 1_3^t \otimes Z(9) \\ -1_{18} & 1_9 \otimes I_2 & 1_3 \otimes Z(9) & Z(27) \end{bmatrix}, \tag{16}$$

with

$$Z(27) = (C_3^{-1} \otimes I_9)C_{27}. \tag{17}$$

Writing

$$C_{27} = \begin{pmatrix} X(27) & X^*(27) \\ X^*(27) & X(27) \end{pmatrix}, \tag{18}$$

we have

$$Z(27) = \begin{pmatrix} Z_1(27) & Z_2(27) \\ Z_2(27) & Z_1(27) \end{pmatrix}, \tag{19}$$

where

$$Z_1(27) = vX(27) - v^2 X^*(27), \tag{20}$$

and

$$Z_2(27) = vX^*(27) - v^2 X(27), \qquad v = e^{2\pi i/3}. \tag{21}$$

We can see that matrix A is a real matrix.

　　Set

$$H = 1 \oplus F(2) \oplus (F(2) \otimes I_3) \oplus (F(2) \otimes I_9), \qquad (22)$$

and

$$B = H^{-1}AH^{-1}. \qquad (23)$$

We can write

$$F_\pi = CHBH. \qquad (24)$$

We also have the variant

$$F_\pi = HDBH, \qquad (25)$$

where

$$D = H^{-1}CH. \qquad (26)$$

D is the diagonal matrix

$$D = 1 \oplus D_3 \oplus (D_3 \otimes I_3) \oplus (D_3 \otimes I_9), \qquad (27)$$

with

$$D_3 = \begin{bmatrix} 1 & & \\ & v & \\ & & v^2 \end{bmatrix}, \quad v = exp(2\pi i/3).$$

The arithmetic count of (25) is given in table 4.1.

Tables of Arithmetic Counts

Table 2.1. Algorithm I $F_\pi(9)$

Factor	R.A.	R.M.
C	144	160
A	36	0
$F_c(9)$	180	160
$F_\Pi(9)$	204	176

R.A.– the number of real additions.

R.M.– the number of real multiplications.

Table 2.2. Algorithm II $F_\pi(9)$

Factor	R.A.	R.M.
H	16	0
T	0	8
$F''_\pi(9)$	56	36+{8}
$F_\pi(9)$	88	44+{8}

Table 4.1. $F_\pi = HDBH$ (N=27)

Factor	R.A.	R.M.
H	52	0
D	0	26
B	450	414
F_π	554	440

[References]

[1] Blahut, R. E. *Fast Algorithms for Digital Signal Processing*, Chapter 4 and 8. Addison-Wesley, Reading, Mass., 1985.

[2] Heideman, M. T. *Multiplicative Complexity, Convolution, and the DFT*, Chapter 5. Springer-Verlag, 1988.

[3] Lu, Chao: *Fast Fourier Transform Algorithms For Special N's and The Implementations On VAX*. Ph.D. Dissertation. Jan. 1988, the City University of New York.

[4] Lu, Chao and Tolimieri, R :"Extension of Winograd Multiplicative Algorithm to Transform Size N=p^2q, p^2qr and Their Implementation", Proceeding of ICASSP 89, 19.D.3. Scotland.

[5] Nussbaumer, H. J. *Fast Fourier Transform and Convolution Algorithms*, Second Edition, Springer-Verlag, 1982.

[6] Tolimieri, R. Lu, Chao and Johnson, W. R. :"Modified Winograd FFT Algorithm and Its Variants for Transform Size N=p^k and Their Implementations" accepted for publication by Advances in Applied Mathematics.

[7] Winograd, S. "On Computing the Discrete Fourier Transform", *Proc. Nat. Acad. Sci. USA.*, vol. 73. no. 4,(April 1976):1005-1006.

[8] Winograd, S. "On Computing the Discrete Fourier Transform", *Math. of Computation*, Vol.32, No. 141, (Jan. 1978):pp 175-199.

Problems

1. List all the elements of the unit group $U(25)$ of $Z/25$.

2. Reindex $Z/25$ by its orbits D_0, D_1 and D_2.

3. Write the explicit matrix of $F_\pi(25)$, derive algorithm I and its variants.

4. Find the arithmetic counts for each of the variants derived in problem 3.

5. Derive the algorithm II for $F_\pi(25)$, find its arithmetic counts.

6. Prove the formulas given in table 2 of section 3.

7. Prove the formulas given in table 4 of section 3.

8. Find a generator for unit group of $U(125) = U(5^3)$.

9. Find the system of idempotents $\{e_1, e_2\}$ for $Z/75 = Z/5^23$.

10. Ordering the unit group $U(75)$ using the idempotents found in Problem 9.

11. Derive in detail the algorithm for $F_\pi(75)$ follow the procedures of section 6.

12. Give the arithmetic counts for the algorithm and its variants of 75-point Fourier transform of Problem 11.

13. Derive the Good-Thomas algorithm for 75-point Fourier transform.

14. Compare the arithmetic counts of the results of Problem 12 and Problem 13.

Chapter 13

PERIODIZATION AND DECIMATION

13.1. Introduction

The ring structure of Z/n provides important tools for gaining deep insights into algorithm design. The fundamental partition of the indexing set Z/p^m, a major step in the Rader-Winograd FT algorithm of the preceeding chapter, was based on the unit group $U(p^m)$. We will now examine how the ideal theory of the ring Z/n can be used for algorithm design.

We adopt the point of view that n-point data is a complex-valued function having the set Z/n as its domain of definition. Denote by

$$L(Z/n), \tag{1}$$

the set of all complex-valued functions on Z/n and regard $L(Z/n)$ as a complex vector space under the following rules of addition and scalar multiplication:

$$(I)\ (f + g)(j) = f(j) + g(j) \tag{2}$$

$$(II)\ (af)(j) = a(f(j)),\ f, g \in L(Z/n),\ a \in C,\ 0 \le j < n \tag{3}$$

The ideas of this chapter are best described on the function theoretic level and constitute a part of abelian harmonic analysis. In this section, we will redefine the Fourier transform as a linear operator of the vector space $L(Z/n)$ whose matrix relative to the standard basis, defined below, is the n-point FT $F(n)$.

In the next section, subspaces of $L(Z/n)$ will be introduced corresponding to ideals of Z/n. For such an ideal B, we define the subspace of B-periodic functions and the subspace of B-decimated functions. The main theorem, proved in section 3, establishes an important duality between these subspace determined by the FT. The duality plays a role in both the Cooley-Tukey algorithms and the Rader-Winograd algorithms and provides the key to understanding the global structure of many 1-dimensional and multi-dimensional FT algorithms.

The set of functions

$$\{e_k : 0 \leq k < n\} \tag{4}$$

where

$$e_l(k) = \begin{cases} 1, & k = l \\ 0 & k \neq l \end{cases} \qquad 0 \leq l, k < n \tag{5}$$

is a basis of $L(Z/n)$ called the *standard basis*. If $f \in L(Z/n)$, we can write

$$f = \sum_{k=0}^{n-1} f(k)e_k \tag{6}$$

and we call the n-tuple of components

$$\underline{f} = \begin{pmatrix} f(0) \\ \vdots \\ f(n-1) \end{pmatrix} \tag{7}$$

the *standard representation* of f.

An *inner product* on $L(Z/n)$ is defined by the formula

$$< f, g > = \sum_{j=0}^{n-1} f(j)g^*(j) \tag{8}$$

where $*$ denotes complex conjugation. The standard basis is an *orthonormal basis* relative to this inner product in the sense that

$$< e_j, e_k > = \begin{cases} 1 & j = k \\ 0 & j \neq k \end{cases} \qquad 0 \leq j, k < n \tag{9}$$

The space $L(Z/n)$ viewed as an inner product space with inner product (8) is denoted by $L^2(Z/n)$.

We will define other basis of $L(Z/n)$ bearing some relationship to the ring-structure of Z/n. The first depends solely on the additive qroup structure. The second depends on the product in Z/n.

A function

$$\chi : Z/n \mapsto C^x, \tag{10}$$

C^x the multiplicative group of non-zero complex numbers, is called an additive character if the following condition holds:

$$\chi(l + k) = \chi(l)\chi(k), \qquad 0 \le l, k < n, \tag{11}$$

where the addition in (11) is taken mod n. In mathematical language, an additive character is a homomorphism from the additive group Z/n into the multiplicative group C^x.

The additive characters on Z/n can be described as follows. Denote the subgroup of C^x consisting of all n-th roots of unity by U_n. Then we have the following result.

Theorem 1. An additive character χ of Z/n is a homomorphism of the additive group Z/n into the multiplicative group U_n and is uniquely determined by $\chi(1)$, by the formula

$$\chi(j) = \chi(1)^j, \quad 0 \le j, k < n. \tag{12}$$

The group U_n is a cyclic group having the element

$$\omega = exp(2\pi i/n) \tag{13}$$

as a generator. For each $k, 0 \le k < n$, we define the mapping

$$\chi_k : Z/n \mapsto U_n, \tag{14}$$

by setting

$$\chi_k(j) = \omega^{kj}, \qquad 0 \le j, k < n. \tag{15}$$

By theorem 1, the set,

$$\{\chi_k : 0 \le k < n\} \tag{16}$$

is the set of additive characters on Z/n.

Theorem 2. The set of additive characters on Z/n is an orthogonal basis of $L^2(Z/n)$ and for any additive character χ on Z/n, we have that

$$\|\chi\|^2 = <\chi, \chi> = n \tag{17}$$

Proof If ν is an n-th root of unity, then

$$\sum_{l=0}^{n-1} \nu^l = \begin{cases} 0 & \nu \neq 1 \\ n & \nu = 1 \end{cases}.$$

Take $0 \leq l, k < n$. By (8) and (15), we have that,

$$< \chi_l, \chi_k > = \sum_{r=0}^{n-1} \omega^{r(l-k)} = \begin{cases} 0 & l \neq k \\ n & l = k \end{cases}.$$

This implies that ths set (16) is an orthogonal subset of $L(Z/n)$. Since there are n distinct elements in (16), the same as the dimension of $L(Z/n)$, the set (16) is a basis of $L(Z/n)$, completing the proof of the theorem.

The FT of Z/n can now be defined as the linear operation F of $L(Z/n)$ satisfying the following condition:

$$F(e_j) = \chi_j, \quad 0 \leq j < n \tag{18}$$

By (6), we have that

$$F(f) = \sum_{j=0}^{n-1} f(j)\chi_j, \quad f \in L(Z/n) \tag{19}$$

Setting $g = F(f)$, by (15), we have that

$$g = F(n)\underline{f}, \tag{20}$$

which implies that the matrix of the Fourier transform F of Z/n relative to the standard basis is $F(n)$.

There are useful properties of the linear operator F that will be needed throughout this chapter. They can be proved directly or on the corresponding properties of the matrix $F(n)$. We list them without proof.

$$F^4 = n^2 I, \tag{21}$$

I the identity operator on $L(Z/n)$, and

$$F^2(e_j) = e_{n-j}, \qquad 0 \le j < n \tag{22}$$

$$< f,g >= 1/n < F(f), F(g) >, \qquad f,g \in L(Z/n) \tag{23}$$

13.2. Periodic and Decimated Data

A subset B of Z/n is called an *ideal* if the following two conditions hold:

(I) *B is a subgroup of the additive group Z/n,* $\qquad$ (1)

(II) $\quad B\, Z/n \subset B.$ $\qquad$ (2)

Ideals of Z/n were introduced in chapter 1. where we proved that every ideal B of Z/n has the form

$$B = r\, Z/n, \tag{3}$$

where r is a divisor of n. From (3), we have

$$B = \{rk : 0 \le k < s\}, \quad n = rs \tag{4}$$

Although condition (I) implies condition (II) for the ring Z/n, this is not generally true for other rings and in any case, we want to emphasize condition (II).

The ring structure of Z/n gives rise to the bilinear pairing

$$< l,k >= \omega^{lk}, \quad \omega = e^{2\pi i/n}, \quad 0 \le l,k < n \tag{5}$$

The product, lk, can be taken either *mod n* or in Z since $\omega^n = 1$. Direct computation shows that the bilinear pairing (5) satisfies the following three properties:

$$(I) \quad <l+k,m> = <l,m><k,m>, \tag{6}$$

$$(II) \quad <lk,m> = <l,m>^k, \tag{6}$$

$$(III) \quad <l,k> = <k,l>, \quad 0 \le l,k,m < n \tag{8}$$

The *dual* $B^\perp$ of an ideal B of Z/n is defined by setting

$$B^\perp = \{l \in Z/n : <l,k> = 1, for\ all\ k \in B\}. \tag{9}$$

Conditions (6) and (7) imply that $B^\perp$ is also an ideal.

Example 1. Take $n = 6$ and $B = 2Z/6$. Then

$$B^\perp = 3Z/6$$

Example 2. Take $n = 9$ and $B = 3Z/9$, Then $B^\perp = B$.

Theorem 1. If $B = rZ/n$ then $B^\perp = sZ/n$, $n = r.s$

Proof Since $n = r \cdot s$ and $\omega^n = 1$, we have that

$$B^\perp \supset sZ/n.$$

Conversely, if $k \in B^\perp$ then, $\omega^{kr} = 1$ which implies that s divides k and

$$B^\perp \subset sZ/n,$$

completing the proof of the theorem.

An immediate consequence of theorem 1 is that

$$(B^\perp)^\perp = B \tag{10}$$

Take $B = rZ/n$, $n = r \cdot s$. A function $f \in L(A)$ is called B-*periodic* if the following condition is satisfied;

$$f(a + b) = f(a), \qquad a \in A, b \in B. \tag{11}$$

A B-periodic function f is uniquely determined by the vector of values

$$\underline{g} = \begin{pmatrix} f(0) \\ \vdots \\ f(r-1) \end{pmatrix}, \tag{12}$$

by the formula,

$$f(l + rk) = f(l), \quad 0 \le l < r, \ 0 \le ks \tag{13}$$

In vector notation

$$\underline{f} = 1_s \otimes \underline{g}, \tag{14}$$

where $\underline{f}$ is the standard representation of f.

Take $g \in L(Z/r)$ and define

$$f = \pi^*(g) \in L(Z/n) \tag{15}$$

by the formula

$$f(l + kr) = g(l), \quad 0 \le l < r, 0 \le k < s. \tag{16}$$

Observe that f is B-periodic. By (13) every B-periodic function is of this form and we have the following result.

Theorem 2. The mapping

$$\pi^* : L(Z/r) \mapsto L(Z/n),$$

is a linear-isomorphism of $L(Z/r)$ onto the space of B-periodic functions in $L(Z/n)$, $B = rZ/n$.

Define

$$\pi : Z/n \mapsto Z/r \tag{17}$$

by setting $\pi(x) = x'$ where $x \cong x' \bmod r$ and $0 \le x' < r$. The mapping π is a ring-homomorphism of Z/n onto Z/r.

$$B \equiv rZ/n = \{a \in Z/n : \quad \pi(a) = 0\}. \tag{18}$$

Using (15), the linear isomorphism π^* can be described by the formula

$$\pi^*(g) = g \cdot \pi, \qquad g \in L(Z/r). \tag{19}$$

A function $f \in L(Z/n)$ is called B-*decimated* if we have

$$f(a) = 0, \quad a \notin B. \tag{20}$$

If f is B-decimated then

$$\underline{f} = \begin{pmatrix} f(0) \\ f(r) \\ \vdots \\ f(n-1) \end{pmatrix} \otimes \underline{d}, \tag{21}$$

where $\underline{d}$ is the r-tuple

$$\underline{d} = \begin{pmatrix} 1 \\ 0 \\ \vdots \\ 0 \end{pmatrix}. \tag{22}$$

Take $h \in L(Z/s)$ and define

$$f = \sigma^*(h) \in L(Z/n),$$

by setting

$$f(kr) = h(k) \quad 0 \le k < s, \tag{23}$$

and f equal to 0 otherwise. Then we have the next result.

Theorem 3.

$$\sigma^* : L(Z/s) \mapsto L(Z/n) \tag{24}$$

is a linear-isomorphism of $L(Z/s)$ onto the B-decimated functions in $L(Z/n)$.

13.3. <u>FT of Periodic and Decimated Data</u>

We come now to the first major result of this chapter; the duality between spaces of periodic functions and spaces of decimated functions determined by the Fourier transform. The duality is the function theoretic analog of theorem 1. of section 2.

Take $B = rZ/n, n = r \cdot s$. Suppose $f \in L(Z/n)$ is B-periodic and $F(f)$ is its Fourier transform. By definition

$$F(f)(l) = \sum_{k=0}^{n-1} f(k) < k,l >, \qquad 0 \le l < n. \tag{1}$$

Using (4) of section 2, we see that every k, $0 \le k < n$, can be written uniquely in the form,

$$k = k' + b, \qquad 0 \le k < r, \ b \in B. \tag{2}$$

It follows that

$$F(f)(l) = \sum_{k'=0}^{r-1} \sum_{b \in B} f(k' + b) < k',l > < b,l >, 0 \le k < n \tag{3}$$

implying by B-periodicity that

$$F(f)(l) = \sum_{k=0}^{r-1} f(k) < k,l > \cdot \sum_{b \in B} < b,l >, 0 \le l < n. \tag{4}$$

We want to compute

$$\gamma(l) = \sum_{b \in B} < b,l >, \qquad 0 \le l < n \tag{5}$$

There are two cases to consider. First suppose $l \in B^{\perp}$. Then, by definition

$$< b,l > = 1, \quad b \in B \tag{6}$$

and we have that

$$\gamma(l) = s, \quad l \in B^{\perp}. \tag{7}$$

Otherwise $l \notin B^{\perp}$ and there exists $c \in B$ such that

$$< c, l > \neq 1. \tag{8}$$

Then, by (7) of section 2

$$< c, l > \gamma(l) = \sum_{b \in B} < c + b, l > = \sum_{b \in B} < b, l > = \gamma(l),$$

which in light of (8) can occur only if

$$\gamma(l) = 0, \quad l \notin B^{\perp}. \tag{9}$$

Placing (7) and (9) into (4), we have that

$$F(f)(l) = 0, \quad l \notin B^{\perp}, \tag{10}$$

$$F(f)(l) = s \cdot \sum_{k=0}^{r-1} f(k) < k, l >, \quad l \in B^{\perp}. \tag{11}$$

By theorem 1 of section 2, $B^{\perp} = sZ/n$ which implies that

$$F(f)(ls) = s \cdot \sum_{k=0}^{r-1} f(k) < k, l >^{s}, \quad 0 \le l < r. \tag{12}$$

Since, $< k, l >^{s} = \nu^{lk}$, where $\nu = \omega^{s} = e^{(2\pi i/r)}$, we can rewrite (12) as follows:

$$F(f)(ls) = s \cdot \sum_{k=0}^{r-1} f(k)\nu^{kl}, \quad \nu = e^{(2\pi i/r)} \tag{13}$$

This discussion leads to the next theorem.

Theorem 1. Let $B = rZ/n, n = rs$. If f is B-periodic then $F(f)$ is $B^{\perp}$-decimated, and on $B^{\perp}$, is given by

$$\begin{pmatrix} F(f)(0) \\ F(f)(s) \\ \vdots \\ F(f)(n-s) \end{pmatrix} = s \cdot F(r) \begin{pmatrix} f(0) \\ f(1) \\ \vdots \\ f(r-1) \end{pmatrix}. \tag{14}$$

Observe that computing the n-point FFT of B-decimated data can be carried out using one r-point FFT.

Since the space of B-periodic functions and the space of $B^{\perp}$-decimated functions have the same dimension r. Theorem 1 implies the next result.

Theorem 2. The FT F of Z/n maps the space of B-periodic functions isomorphically onto the space of $B^{\perp}$-decimated functions.

Suppose now that we begin with a B-decimated function $g \in L(Z/n)$. We are still assuming that $B = rZ/n$. By definition, the Fourier transform $F(g)$ of g is given by the formula

$$F(g)(l) = \sum_{k \in B} g(k) < k, l >, \quad 0 \le l < n. \tag{16}$$

Since $l \in B^{\perp}$ implies $< k, l >= 1$, for all $k \in B$, replacing l by $l + s$ in (16) and using (7) of section 2, proves that

$$F(g)(l + s) = F(g)(l), \quad 0 \le l < n, \tag{17}$$

and $F(g)$ is $B^{\perp}$-periodic. The summation in (16) can be taken as indicated in (4) of section 2 and we can rewrite (16) as follows:

$$F(g)(l) = \sum_{k=0}^{s-1} g(rk)\nu^{lk}, \ \nu = \omega^r = e^{(2\pi i/s)}, \ 0 \le l < s, \tag{18}$$

which proves the next result.

Theorem 3. Let $B = rZ/n, n = r \cdot s$. If g is B-decimated then $F(g)$ is $B^{\perp}$-periodic and is given by the matrix formula

$$\begin{pmatrix} F(g)(0) \\ F(g)(1) \\ \vdots \\ F(g)(s-1) \end{pmatrix} = F(s) \begin{pmatrix} g(0) \\ g(r) \\ \vdots \\ g(n-r) \end{pmatrix}. \tag{19}$$

Theorem 2 and 3 express the duality on the function theoretic determined by the Fourier transform. They will serve to give the global structure of several "multiplicative " Fourier transform algorithms. We take up this topic in detail in the following chapters.

Theorem 3 can be viewed as the first step in the n-point Fourier transform Cooley-Tukey algorithm corresponding to the divisor r. Infact, it is a part of the usual formula used to derive this algorithm. We have chosen to break up this formula into what we consider to be its essential components, the main one being formula (19). Observe that formula (19) leads to an algorithm computing the n-point Fourier transform of B-decimated data using the s-point Fourier transform.

13.4. The Ring Z/p^m

The result of the preceeding section will now be applied to the special case of $N = p^m$, $m > 1$ and p an odd prime. The ring

$$A = Z/p^m, \tag{1}$$

is a *local ring* in the sense that it has a *unique maximal ideal*

$$M = pZ/p^m. \tag{2}$$

In fact, by the results of section 3, the ring A has exactly m ideals

$$M^k = p^k A, \quad 1 \le k \le m, \tag{3}$$

and we have

$$(0) = M^m \subset M^{m-1} \subset \cdots \subset M^2 \subset M \subset A. \tag{4}$$

Denote by

$$L(k,l), \quad 0 \le k,l \le m, \tag{5}$$

the subspace of all $f \in L(Z/p^m)$ satisfying the following two conditions:

$$(I) \quad f \text{ is } M^k - decimated, \tag{6}$$

$$(II) \quad f \ is \ M^l - periodic. \tag{7}$$

In particular, $L(1, m)$ is the subspace of all M-decimated functions and $L(0, 1)$ is the subspace of all M-periodic functions.

The dual of the ideal M^j is M^{m-k} by theorem 1 of section 2. Applying theorem 2 and 3 of section 3, we have the following result.

Theorem 1. The Fourier transform of Z/p^m maps the subspace $L(k, l)$ onto the subspace $L(m - l, m - k)$, $0 \le k \le l \le m$.

The subspace $L(k, l)$, $0 \le k \le l \le m$, will play an important part in the overall design of algorithms in the next two chapters. Especially important is the subspace $L(1, m - 1)$, which by theorem 1, is invariant under the Fourier transform F of Z/p^m, in the sense that

$$F(L(1, m - 1)) = L(1, m - 1). \tag{8}$$

Applying (28) of section 1, we see that the linear operator $n^{-1/2}F$ is an unitary operator of $L^2(Z/p^m)$. Denote the orthogonal complement of $L(1, m - 1)$ in $L^2(Z/p^m)$ by W. Then

$$W = \{f \in L^2(Z/p^m) : \ < f, g >= 0, \ for \ all \ g \in L(1, m - 1)\} \tag{9}$$

satisfies the following two properties:

$$(I) \quad L^2(Z/p^m) = W \oplus L(1, m - 1), \tag{10}$$

$$(II) \quad F(W) = W. \tag{11}$$

Properties (8) and (10-11) imply that we can study the action of F on $L(Z/p^m)$ by studying the action of F on $L(1, m - 1)$ and the action of F on W.

To begin with, we construct a basis of the space $L(0, m - 1)$ of M^{m-1}-periodic functions and a basis of $L(1, m - 1)$.

Define the set of functions

$$\{E_k : \ 0 \le kpm - 1\}, \tag{12}$$

by

$$\begin{bmatrix} E_0 \\ E_1 \\ \vdots \\ E_{t-1} \end{bmatrix} = (1_p^t \otimes I_t) \begin{bmatrix} e_0 \\ e_1 \\ \vdots \\ e_{n-1} \end{bmatrix}, \tag{13}$$

where $n = p^m$, $t = p^{m-1}$, is a basis of $L(0, m-1)$. It follows that the set

$$\{E_{pk} : \ 0 \le k < p^{m-2}\} \tag{14}$$

is a basis of $L(1, m-1)$ and if $f \in L(1, m-1)$, then,

$$f = \sum_{k=0}^{s-1} f(pk) E_{pk}. \tag{15}$$

Theorem 2. Let $n = p^m$, $m \ge 2$ and p an odd prime, then the matrix of the restriction of F to $L(1, m-1)$ is $pF(p^{m-2})$, relative to the basis (14).

Proof The function E_{pk}, $0 \le k < s = p^{m-2}$, is equal to 1 on the set

$$S = \{pk + rp^{m-1} : \ 0 \le r < p\}$$

and vanishes otherwise. Let $g = F(E_{pk})$. Then by (8) and (15)

$$g = \sum_{l=0}^{s-1} g(pl) E_{pl}, \quad s = p^{m-2}$$

By definition

$$g(pl) = \sum_{u \in S} w^{upk},$$

$$g(pl) = \sum_{r=0}^{p-1} w^{(pk+rp^{m-1})pl},$$

$$g(pk) = pv^{kl}, \quad 0 \le k, l < s,$$

where $v = w^{p^2} = exp(2\pi i/p^{m-2})$. It follows that

$$g = p \sum_{l=0}^{s-1} v^{kl} E_{pl},$$

completely proving the theorem.

Problems

1. If χ is an additive character of Z/n, then $\chi(1)$ is a n-th root of unity. (without relying on theorem 1).

2. Prove that the set of additive characters of Z/n is a group under the product rule:

$$(\chi\chi')(k) = \chi(k)\chi'(k), \quad k \in Z/n,$$

where χ and χ' additive characters of Z/n.

3. Prove formula (1.20)

4. Prove formula (1.22) and describe the matrix of F^2 relative to the standard basis.

5. Prove that the dual is an ideal.

6. Write down the collection of all B-periodic functions on Z/n where $n = 6$ and $B = 2Z/6$.

7. Repeat Problem 6, where $n = 27$ and $B = 3Z/27$.

8. Write down the collection of all B-decimated functions on Z/n where $n = 6$ and $B = 3Z/6$.

9. Repeat Problem 8, where $n = 27$ and $B = 9Z/27$.

10. Verify $F(L(1, m - 1)) = L(1, m - 1)$ when $n = 3^3$.

Chapter 14

MULTIPLICATIVE CHARACTERS AND THE FFT

14.1. Introduction

Fix an odd prime p throughout this chapter, and set $U(m) \equiv U(Z/p^m)$, the unit group of Z/p^m. Consider the space $L(Z/p^m)$. For $m > 1$, we defined the space

$$L_0 = L(1, m-1), \tag{1}$$

of M-decimated and M^{m-1}-periodic functions on Z/p^m with $M = pZ/p^m$ and proved that

$$L(Z/p^m) = W \oplus L_0, \tag{2}$$

where W is the orthogonal complement of L_0 in $L(Z/p^m)$. The space L_0 and W are invariant under the action of the Fourier transform F of Z/p^m. The action of F on L_0 was described in the preceeding chapter. We will now take up the action of F on W. For this purpose, we introduce the multiplicative characters on the ring Z/p^m.

First consider the field Z/p. A *multiplicative character* x of Z/p is a homomorphism from the multiplicative group $U(1)$ into the multiplicative group $\mathcal{C}^\times$ of non-zero complex numbers :

$$x(ab) = x(a)x(b), \quad a, b \in U. \tag{3}$$

The product ab is taken in $U(1)$ and is multiplication *mod p*.

<u>Example</u> 1. Take $p = 7$. The unit group $U \equiv U(1)$ of $Z/7$ is cyclic of order 6 having $z = 3$ as a generator. We order U exponentially relative to generator $z = 3$,

$$1, 3, 2, 6, 4, 5.$$

Consider any multiplicative character x of $Z/7$. Then x is completely determined by its value on the generator $z = 3$ by the formula

$$x\left(3^k\right) = x(3)^k, \ 0 \le k < 6.$$

In particular, since $3^6 \equiv 1 \ mod \ 7$,

$$x\left(3\right)^6 \equiv x\left(3^6\right) = x(1) = 1,$$

and $x(3)$ is a 6-th root of unity. As a consequence, there exists exactly 6 multiplicative characters of $Z/7$ which can be defined by the following table,

$$\begin{pmatrix} & 1 & 3 & 2 & 6 & 4 & 5 \\ x_0 & 1 & 1 & 1 & 1 & 1 & 1 \\ x_1 & 1 & v & v^2 & v^3 & v^4 & v^5 \\ x_2 & 1 & v^2 & v^4 & v^6 & v^8 & v^{10} \\ x_3 & 1 & v^3 & v^6 & v^9 & v^{12} & v^{15} \\ x_4 & 1 & v^4 & v^8 & v^{12} & v^{16} & v^{20} \\ x_5 & 1 & v^5 & v^{10} & v^{15} & v^{20} & v^{25} \end{pmatrix}$$

where $v = exp(2\pi i/6)$.

In general, the unit group $U \equiv U(1)$ of Z/p is a cyclic group of order $t = p - 1$. Take a generator z of U and order U exponentially relative to z ,

$$1, z, \dots, z^{t-1}. \tag{4}$$

Arguing as in the example, a multiplicative character x of Z/p is completely determined by its value on the generator z by the formula

$$x\left(z\right)^k = x(z^k), \quad 0 \le k < t, \tag{5}$$

and $x(z)$ is a t-root of unity. It follows that to define a multiplicative character x of Z/p we set

$$x\left(z\right) = v^l, \quad 0 \le l < t, \tag{6}$$

where $v = exp\,(2\pi i/t)$ and use formula (5) to extend the definition of x to all of U. In this way, we see that there exists exactly t multiplicative characters of Z/p given by the following table,

Table 1. Multiplicative Characters: Z/p

$$
\begin{pmatrix}
 & 1 & z & z^2 & \cdots & \cdots & z^{t-1} \\
x_0 & 1 & 1 & 1 & \cdots & \cdots & 1 \\
x_1 & 1 & v & v^2 & \cdots & \cdots & v^{t-1} \\
\vdots & \vdots & & & & & \\
\vdots & \vdots & & & & & \\
x_{t-1} & 1 & v^{t-1} & v^{2(t-1)} & \cdots & \cdots & v^{(t-1)(t-1)}
\end{pmatrix}
$$

where $v = exp(2\pi i/t)$. The corresponding formula is

$$x_k\left(z^l\right) = v^{kl}, \quad 0 \le k,l < t. \tag{7}$$

Denote the set of multiplicative characters of Z/p by $\hat{U}(1)$. By (7),

$$\hat{U}(1) = \left\{x_k \mid 0 \le k < t\right\}, \tag{8}$$

which we can view as a vector of functions of size t.

We extend the domain of definition of a multiplicative character x of Z/p to all of Z/p by setting

$$x(0) = 0. \tag{9}$$

In general, the N-point Fourier transform matrix $F(N)$ maps the standard basis of $L(Z/N)$ onto the basis of additive characters. If $N = p$, we have a multiplicative analog. Set

$$E^\times(1) = \left\{e_{z^k} \mid 0 \le k < t\right\}, \subset L(Z/p), \tag{10}$$

and view $E^\times(1)$ as a vector of functions of size t. By table 1,

$$\hat{U}(1) = F(t)E^\times(1). \tag{11}$$

The prime case theory extends in a straightfoward fashion to the general case $N = p^m$, $m \geq 1$. Set $U \equiv U(m)$. Since p is an odd prime, U is a cyclic group of order

$$t = p^{m-1}(p-1). \tag{12}$$

Fix a generator z of U. The exponential ordering of U is given by

$$1, z, z^2, \ldots, z^{t-1}, \tag{13}$$

with $z^t = 1$. It is natural to describe multiplicative characters of Z/p^m relative to the exponential ordering. First a multiplicative character of Z/p^m is a homomorphism of U into $\mathbb{C}^\times$. Arguing as above, there are exactly t multiplicative characters of Z/p^m

$$x_k, \quad 0 \leq k < t, \tag{14}$$

defined by the formula

$$x_k(z^l) = v^{kl}, 0 \leq l < t, \tag{15}$$

where $v = exp(2\pi i/t)$. This leads to the following table.

Table 2. Multiplicative Characters: Z/p^m

	1	z	z^2	... ...	z^{t-1}
x_0	1	1	1	... ...	1
x_1	1	v	v^2	... ...	v^{t-1}
$\vdots$	$\vdots$				
x_{t-1}	1	v^{t-1}	$v^{2(t-1)}$	... ...	$v^{(t-1)(t-1)}$

where $v = exp(2\pi i/t)$ and $t = p^{m-1}(p-1)$.

Denote the set of multiplicative characters of Z/p^m by $\hat{U}(m)$. Relative to the generator z

$$\hat{U}(m) = \{x_k \mid 0 \leq k < t\}. \tag{16}$$

We can view $\hat{U}(m)$ as a vector of functions of size t.

Extend the domain of definition of a multiplicative character x of Z/p^m by setting

$$x(a) = 0, \quad a \notin U. \tag{17}$$

Define

$$E^{\times}(m) = \{\, e_{z^k} \mid 0 \leq k < t \,\} \subset L(Z/P^m), \tag{18}$$

By table 2,

$$\hat{U}(m) = F(t)E^{\times}(m). \tag{19}$$

Take $m \geq 1$. A group multiplication is placed on $\hat{U}(m)$ by the rule

$$x(x'(a)) = x(a)x'(a), \quad a \in U(m), \quad x, \, x' \in \hat{U}(m). \tag{20}$$

We see that

$$x_k x_l = x_{k+l}, \tag{21}$$

with $k + l$ taken $mod\ t$. This implies that $\hat{U}(m)$ is group-isomorphic to $U(m)$. The identity element of $\hat{U}(m)$ is x_0 which is called the *principal character*.

There are several important formulas involving multiplicative characters which will be repeatedly used throughout the work. Take a multiplicative character x of Z/p^m, $m \geq 1$. There are two cases to consider. If $x = x_0$, the principal character then

$$\sum_{k \in U(m)} x_0(k) = t.$$

Otherwise

$$\sum_{k \in U(m)} x(k) = 0, \quad x \neq x_0. \tag{22}$$

To see this, take $k_0 \in U(m)$ such that

$$x(k_0) \neq 1. \tag{23}$$

As k runs over $U(m)$, $k_0 k$ runs over $U(m)$. Then

$$\sum_{k \in U(m)} x(k_0\, k) = \sum_{k \in U(m)} x(k). \tag{24}$$

Since

$$x(k_0 k) = x(k_0)x(k), \tag{25}$$

we also have

$$\sum_{k \in U(m)} x(k_0 k) = x(k_0) \sum_{k \in U(m)} x(k), \tag{26}$$

which in light of (23) and (24), implies the result.

Two additional formulas will also be useful in the following sections. For two distinct multiplicative characters x and x' of Z/p^m,

$$< x, x' > = \sum_{k \in U(m)} x(k)(x')^*(k) = 0. \tag{27}$$

We also have

$$< x, x > = o(U(m)) = t. \tag{28}$$

Consequently, $\hat{U}(m)$, the set of multiplicative characters of Z/p^m is an *orthogonal subset* of the inner product space $L^2(Z/p^m)$.

14.2. Periodicity

2.1 Periodic multiplicative characters

Subspaces of $L(Z/p^m)$ will be constructed from sets of multiplicative characters satisfying periodicity conditions and their FT's. If $m = 1$ then Z/p is a field and has no non-trivial ideals. Periodicity plays no role. However, two subsets of the set $\hat{U}(1)$ of multiplicative characters of Z/p will be distinguished. Denote by $V(1)$ the set difference

$$V(1) = \hat{U}(1) - \{x_0\}. \tag{1}$$

Let

$$V_0(1) = \{x_0\}. \tag{2}$$

A multiplicative character $x \in V(1)$ is called *primitive*. From section 1,

$$V(1) = M(1)\, E^\times(1), \tag{3}$$

where $M(1)$ is the matrix formed by erasing the first row of $F(p-1)$.

Assume now that $m > 1$.

$$M = pZ/p^m, \tag{4}$$

is the unique maximal ideal of the ring Z/p^m. Set

$$\hat{U}_k(m) = \{x \in \hat{U}(1) \mid x \text{ is } M^k - periodic\}, \quad 0 \le k \le m. \tag{5}$$

$\hat{U}_0(m)$ is the empty set and $\hat{U}(m) = \hat{U}_m(m)$. Also $\hat{U}_k(m), 1 \le k \le m$, is a subgroup of $\hat{U}(m)$ where

$$\hat{U}_{k-1}(m) \subset \hat{U}_k(m). \tag{6}$$

Form the set diferrences

$$V_k(m) = \hat{U}_k(m) - \hat{U}_{k-1}(m), \quad k \ge 2, \tag{7}$$

$$V_1(m) = \hat{U}_1(m) - \{x_0\}, \tag{8}$$

$$V_0(m) = \{x_0\}. \tag{9}$$

The set

$$V(m) = V_m(m) \tag{10}$$

is called the *set of primitive multiplicative characters* on Z/p^m. $\hat{U}(m)$ is the disjoint union

$$\hat{U}(m) = \cup_{k=0}^m V_k(m). \tag{11}$$

Example 1. Relative to the generator $z=2$ of the unit group U of $Z/9$,

$$\hat{U}_1(2) = \{x_0, x_3\},$$

$$V(2) = \{x_1, x_2, x_4, x_5\}.$$

Example 2. Relative to the generator $z=2$ of the unit group U of $Z/27$,

$$\hat{U}_1(3) = \{x_0, x_9\},$$

$$\hat{U}_2(3) = \{x_0, x_3, x_6, x_9, x_{12}, x_{15}\},$$

$$V(3) = \{x_k \in \hat{U}(3) \mid 0 \le k < 18, \ (3, k) = 1\}.$$

Consider first the case $m = 2$. Take a generator z of the unit group $U = U(2)$ of Z/p^2. Set

$$t = o(U) = p(p - 1).$$

Form the p by $p - 1$ array

$$\begin{pmatrix} 1 & z & \cdot & \cdot & \cdot & z^{p-2} \\ z^{p-1} & & & & & \vdots \\ \vdots & & & & & \vdots \\ z^{(p-1)^2} & & & & & z^{t-1} \end{pmatrix}. \tag{19}$$

The main point is that the exponential ordering is compatible with ideal periodicity in the sense that two elements of U are equal *mod p* if and only if they lie in the same column of the array. It is sufficient to show that

$$z^{p-1} \equiv 1 \ mod \ p, \tag{20}$$

which follows from the fact that z mod p generates the unit group $U(p)$ of Z/p and $U(p)$ has order $p - 1$.

Take a multiplicative character x_k of Z/p^2. Suppose x_k is M-periodic where

$$M = pZ/p^2. \tag{21}$$

Then x_k takes the same value at each point of a column of the array (12). In particular

$$x_k(z^{p-1}) = v^{k(p-1)} = 1, \tag{22}$$

where $v = exp\,(2\pi i/t)$. It follows that

$$p\,/\,k. \tag{23}$$

Conversely condition (12) implies x_k is M-periodic, proving the following theorem.

Theorem 1. Relative to any generator z of the unit group $U(2)$ of Z/p^2,

$$\hat{U}_1(2) = \{x_k \mid 0 \le k < t, p/k\}, \tag{24}$$

$$V(2) = \{x_k \in \hat{U}(2) \mid 0 \le k < t,\ (p,k) = 1\}. \tag{25}$$

Consider the general case $m > 1$. Fix a generator z of $U(m)$, for the rest of this section. We will describe the subgroup $\hat{U}_k(m)$ of all M^k-periodic multiplicative characters on Z/p^m, where $M = pZ/p^m$. First

$$z_k \equiv z\ mod\ p^k$$

generates $U(k)$. Set $t_k = o(U(k)) = p^{k-1}(p-1)$. Since $z_k^{t_k} = 1$ on $U(k)$,

$$z^{t_k} \equiv 1\ mod\ p^k.$$

Moreover, t_k is the smallest positive power of z having this property. Arrange $U(m)$ in the $p^{m-k} \times t_k$ array

$$\begin{bmatrix} 1 & z & \cdot & \cdot & \cdot & z^{t_k-1} \\ z^{t_k} & & & & & \\ \cdot & & & & \cdot & \\ \cdot & & & & \cdot & \\ \cdot & & & & \cdot & \\ z^{(q-1)t_k} & & \cdot & \cdot & \cdot & z^{t-1} \end{bmatrix}, \tag{26}$$

where $t = p^{m-1}(p-1)$, and $q = p^{m-k}$. Two elements of $U(m)$ are equal $mod\ p^k$ if and only if they lie in the same column of this array. Take $x_l \in \hat{U}(m)$. If x_l is M^k-periodic then

$$x_l(z^{t_k}) = v^{lt_k} = x_l(1) = 1, \quad v = exp(2\pi i/t),$$

which is equivalent to

$$p^{m-k}/l. \tag{27}$$

Conversely, condition (27) implies x_l is M^k-periodic, since, in this case

$$x_l(z^r z^{t_k}) = x_l(z^r)x_l(z^{t_k}) = x_l(z^r), \quad 0 \le r < t_k. \tag{28}$$

We have proved the following result.

Theorem 2. Relative to any generator z of the unit group $U(m)$ of Z/p^m,

$$\hat{U}_k(m) = \{x_l \in \hat{U}(m) \mid 0 \le l < t,\ p^{m-k}/l\}. \tag{29}$$

Also

$$V(m) = \{x_l \in \hat{U}(m) \mid 0 \le l < t,\ (p^m, l) = 1\}, \tag{30}$$

$$V_k(m) = \{x_l \in \hat{U}(m) \mid 0 \le l < t,\ (p^m, l) = p^{m-k}\}. \tag{31}$$

2.2 Periodization Map

We will now rewrite the result of section 2.1 using the periodization map of the preceeding chapter. This will permit direct application of the FT formulas derived at that time. Consider the ring-homomorphism

$$\pi_k : Z/p^m \longrightarrow Z/p^k, \tag{32}$$

defined by $\pi_k(a) \equiv a \bmod p^k$. The induced *periodization map*

$$\pi_k^* : L(Z/p^m) \longrightarrow L(Z/p^k), \tag{33}$$

is given by $\pi_k^*(f) = f \circ \pi_k$. We proved in chapter 13 that

$$\pi_k^*(L(Z/p^k)) \tag{34}$$

is the subspace of all M^k-periodic functions in $L(Z/p^k)$. Since π_k restricts to a group-homomorphism of $U(m)$ onto $U(k)$, we have the next result.

Theorem 3. The periodization map

$$\pi_k^* : L(Z/p^m) \longrightarrow L(Z/p^k)$$

restricts to a group-isomorphism of $\hat{U}(k)$ onto $\hat{U}_k(m)$ and to a bijection from $V(k)$ onto $V_k(m)$.

Explicitly, if z is a generator of $U(m)$, then π_k maps each column of the array (26) onto a unique element in $U(k)$. Denote the multiplicative characters on Z/p^k relative to $z_k = \pi_k(z)$ by x_l', $0 \leq l < t_l = p^{k-1}(p-1)$. Then

$$x_l'(z_k) = u^l, \quad u = exp(2\pi i/t_k) \tag{35}$$

and

$$x_{lp^{m-k}} = \pi_k^*(x_l'), \tag{36}$$

where

$$x_{lp^{m-k}}(z) = v^{lp^{m-k}}, \quad v = exp(2\pi i/t), \ t = p^{m-1}(p-1). \tag{37}$$

Ideal periodicity can be translated into multiplicative periodicity when applied to multiplicative characters. First the ring homomorphism Π_k restricts to a group-homomorphism of U into $U(k)$ satisfying

$$1 + M^k = \{a \in U \mid \Pi_k(a) = 1\}. \tag{38}$$

Set $G_k = 1 + M^k$. Then x is M^k-periodic if and only if

$$x(G_k) = 1. \tag{39}$$

To see this, take $x \in \hat{U}_k$, $u \in U$ and $m \in M^k$. Then

$$x(u + m) = x(u), \tag{40}$$

$$x(u + m) = x((1 + u^{-1}m)) = x(u)x(1 + u^{-1}m), \tag{41}$$

implying

$$x(1 + u^{-1}m) = 1, \tag{42}$$

proving (27). The converse follows by reversing the steps of the argument.

14.3. FT's of Multiplicative Characters: Z/p

Denote the Fourier transform of Z/p by F. There are two cases to consider depending on whether x is a primitive multiplicative character or $x = x_0$, the principal character.

Theorem 1. Let x_0 be the principal multiplicative character of Z/p. Then

$$F(x_0)(a) = \begin{cases} p - 1 & \text{if} \quad a = 0 \\ -1 & \text{if} \quad 0 < a < p. \end{cases} \tag{1}$$

Proof By definition

$$F(x_0)(a) = \sum_{k=1}^{p-1} v^{ak}, \quad v = exp(2\pi i/p).$$

If $a = 0$, then the sum on the right equals $p - 1$. If $0 < a < p$ then since

$$\sum_{k=0}^{p-1} v^{ak} = -1,$$

the sum equals -1, completing the proof.

Theorem 2. If x is a primitive multiplicative character of Z/p then

$$F(x) = G_1(x) x^*, \qquad (2)$$

where $G_1(x)$ is the complex constant

$$G_1(x) = F(x)(1) = \sum_{k=1}^{p-1} x(k) v^k. \qquad (3)$$

Proof By definition

$$F(x)(a) = \sum_{k=1}^{p-1} x(k) v^{ak}.$$

If $a = 0$ then

$$F(x)(0) = \sum_{k=0}^{p-1} x(k) = 0.$$

Suppose now $0 < a < p$. Then a is invertible in Z/p and we can write

$$\sum_{k \in U} x(k) v^{ak} = \sum_{k \in U} x(a^{-1}k) v^k = \left(\sum_{k \in U} x(k) v^k \right) x(a^{-1}).$$

Since $x(a^{-1}) = x(a)^*$, the theorem follows.

The complex constant $G_1(x)$ is called a *Gauss sum*. Since $p^{1/2} F$ is unitary, if x is primitive, we have

$$< F(x), F(x) >= p < x, x, \qquad (4)$$

and by (2)

$$< F(x), F(x) >= |G_1(x)|^2 < x^*, x^* >, \qquad (5)$$

implying

$$|G_1(x)|^2 = p, \quad x \ primitive. \qquad (6)$$

Example 1. If $p = 3$ then

$$F\left(x_0\right)(a) = \begin{cases} 2, & a = 0 \\ -1, & a = 1,2 \end{cases},$$

$$F\left(x_1\right) = G_1\left(x_1\right) x_1,$$

where

$$G_1\left(x_1\right) = w - w^2, \quad w = exp\left(2\pi i/3\right).$$

Example 2. If $p = 5$ then

$$F\left(x_0\right)(a) = \begin{cases} 4, & a = 0 \\ -1, & a = 1,2,3,4 \end{cases}.$$

$$F\left(x_0\right) = G_1\left(x_1\right) x^*,$$

where the Gauss sums $G_k = G_1\left(x_k\right)$ are given by

$$G_1 = w - w^2 + i\left(w^2 - w^5\right),$$

$$G_2 = w + w^4 - \left(w^2 + w^3\right),$$

$$G_3 = \left(w - w^4\right) - i\left(w^2 - w^3\right), \quad w = exp\left(2\pi i/5\right).$$

14.4. **FT's of Multiplicative Characters** Z/p^m

Take $m > 1$. First we compute the Fourier transforms of primitive multiplicative characters of Z/p^m.

Theorem 1. If $x \in V(m)$ then

$$F\left(x\right) = G_m\left(x\right) x^*, \tag{1}$$

where $G_m(x)$ is the complex constant

$$G_m(x) = F\left(x\right)(1). \tag{2}$$

Proof If $a \in U$, the unit group of Z/p^m, then arguing exactly as in the preceeding section

$$F\left(x\right)\left(a\right) = G_m\left(x\right) x^*\left(a\right), \ a \in U.$$

Assume now that $a \notin U$. Write

$$a = p\,a', \ 0 \leq a' < p^{m-1}.$$

Since x is not M^{m-1}-periodic, by (2.40).

$$x\left(c\right) \neq 1,$$

for some $c \in 1 + M^{m-1}$. Observe $p \equiv pc \ mod \ p^m$. Let $w = exp(2\pi i/p^m)$, then

$$w^{ab} = w^{pba'c} = w^{bac}, \quad b \in U,$$

and we have

$$F\left(x\right)\left(a\right) = \sum_{b \in U} x\left(b\right) w^{abc} = \sum_{b \in U} x(c^{-1} b)\, w^{ab} = x\left(c\right)^{\times} F(x)\left(a\right),$$

which since $x\left(c\right) \neq 1$ implies

$$F\left(x\right)\left(a\right) = 0,$$

proving the theorem.

Formulas computing FT's of *non-primitive multiplicative characters* will now be derived. Take for this discussion

$$V_k \equiv V_k(m), \quad x \in V_j, \quad 0 \leq k < m. \tag{3}$$

The principal multiplicative character will be handled separately. Since x is M^k-periodic, the FT of x is M^{m-k}-decimated and vanishes off of the set

$$M^{m-k} = p^{m-k} \ Z/p^m. \tag{4}$$

Applying results from preceeding section

$$x = \pi_k^*(\psi), \tag{5}$$

where ψ is a primitive multiplicative character of Z/p^k. Then $F(x)$ can be computed by decimating the FT $F'(\psi)$ to M^{m-k}. The exact formula will be given below. Define a function $g_x \in Z/p^m$ by setting

$$g_x\left(p^{m-k}\,u\right) = x\left(u\right), \ u \in U, \tag{6}$$

and setting g_x equal to zero otherwise. Formula (16) is well-defined since x is M^k-periodic.

Theorem 2. If $x \in V_k$, $1 \le k < m$, then

$$F(x) = p^{m-k}G_k(x)g_{x^*}, \tag{7}$$

where $G_k(x)$ is the complex constant

$$G_k\left(x\right) = G_k\left(\psi\right). \tag{8}$$

Proof By the preceeding chapter

$$F(x)(p^{m-k}l) = p^{m-k}\,F'(\psi)(l), \ 0 \le l < p^k.$$

Applying theorem 1.

$$F'\left(\psi\right)\left(l\right) = G_k\left(\psi\right)\psi^*(l), \ 0 \le l < p^k.$$

Since

$$\psi\left(l\right) = x(l), \ 0 \le l < p^k,$$

the theorem follows.

Set

$$V_k' = \left\{\, g_x \,:\, x \in V_k \,\right\}. \tag{9}$$

Relative to a fixed generator z of U we use the abbreviation

$$g_l \equiv g_{z_l}, \quad 0 \leq l < t, \quad (p^m, l) = p^k. \tag{10}$$

Example 1. Take $z = 2$ as a generator of the unit group U of $Z/9$. Then

$$V_1' = \{ g_3 \},$$

where g_3 vanishes off of the set

$$3\,U = \{ 3, 6 \}$$

and on this set it takes the values

$$\begin{pmatrix} & 3 & 6 \\ g_3 & 1 & -1 \end{pmatrix}.$$

Example 2. Take $z = 2$ as a generator of the unit group U of $Z/27$. Then each function in

$$V_2' = \{ g_3, g_6, g_{12}, g_{15} \},$$

vanishes off of the set

$$3\,U = \{ 3, 6, 12, 24, 21, 15 \},$$

and on this set

$$\begin{pmatrix} & 3 & 6 & 12 & 24 & 21 & 12 \\ g_3 & 1 & u & u^2 & u^3 & u^4 & u^5 \\ g_6 & 1 & u^2 & u^4 & u^6 & u^8 & u^{10} \\ g_{12} & 1 & u^4 & u^8 & u^{12} & u^{16} & u^{20} \\ g_{15} & 1 & u^5 & u^{10} & u^{15} & u^{20} & u^{25} \end{pmatrix}.$$

We also have

$$V_1' = \{ g_9 \}$$

where g_9 vanishes off of the set

$$9\,U = \{\,9, 18\,\}$$

and on the set is given by

$$\begin{pmatrix} & 9 & 18 \\ g_9 & 1 & -1 \end{pmatrix}.$$

Consider now the principal multiplicative character x_0 of Z/p^m. Define the function g_0 on Z/p^m by setting g_0 equal to zero off of the set M^{m-1} and setting

$$\begin{pmatrix} & 0 & p^{m-1} & 2p^{m-1} & \cdots & \cdots & (p-1)p^{m-1} \\ g_0 & -(p-1) & 1 & 1 & \cdots & \cdots & 1 \end{pmatrix}. \tag{11}$$

Direct computation shows the following result.

Theorem 3. If x_0 is the principal multiplicative character of Z/p^m then $F(x_0)$ is M^{m-1}-decimated and

$$F(x_0) = -p^{m-1}\,g_0. \tag{12}$$

14.5. Orthogonal Basis Diagonalizing FT $N = p$

Fix a generator z of the unit group U of Z/p. The set of multiplicative characters $\hat{U}$ of Z/p is orthogonal by the formulas discussed in section 2. Set $t = p - 1$ and

$$r = (p-1)\,/\,2. \tag{1}$$

The function e_0 in $L(Z/p)$ taking the value 1 at the point 0 and vanishing on U is orthogonal to $\hat{U}$. It follows that the set

$$e_0,\ x_0,\ x_1, \ldots,\ x_{t-1} \tag{2}$$

is orthogonal and by dimension is a basis of $L^2(Z/p)$.

Set

$$b_0 = \{ e_0, t^{-1/2} x_0 \}, \tag{3}$$

$$b_k = \{ t^{-1/2} x_k, t^{-1/2} p^{-1/2} F(x_k) \}, \; 1 \le k < r, \tag{4}$$

$$b_r = \{ t^{-1/2} x_r \}. \tag{5}$$

The formulas of section 3 imply that $F(x_k)$ is a constant multiple of x_{-k}, $1 \le k < r$, where $-k$ is taken $mod\, t$. As a consequence, the set

$$\bigcup_{k=0}^{r} b_k \tag{6}$$

is an orthogonal subset of $L^2(Z/p)$. In fact, (6) is an *orthogonal basis* by the formulas of section 2. Also, $p^{-1/2} F$ preserves inner products.

Set

$$W_k = Ln\,(b_k), 0 \le k \le r. \tag{7}$$

Consider first the action of F on the subspaces W_k, $1 \le k < r$. Since

$$F^2(f)(a) = Ff(a), \; a \in Z/p, \; f \in L(Z/p), \tag{8}$$

we have

$$F^2(x_k)(a) = p\, x_k(-a) = p\, x_k,(-1)\, x_k\,(a), \; a \in Z/p. \tag{9}$$

This implies that W_k, $1 \le k < r$, is invariant under the action of F and that the matrix of F restricted to W_k relative to the basis b_k is

$$p^{\frac{1}{2}} M_k, \; 1 \le k < r, \tag{10}$$

where

$$M_k = \begin{pmatrix} 0 & x_k(-1) \\ 1 & 0 \end{pmatrix}, \; 1 \le k < r. \tag{11}$$

Using

$$x_k(-1) = (-1)^k, \tag{12}$$

we have

$$M_k = \begin{pmatrix} 0 & (-1)^k \\ 1 & 0 \end{pmatrix}, \; 1 \le k < r. \tag{13}$$

Consider the action of F on W_0. Theorem 1 of section 3 implies that the matrix of F restricted to W_0 relative to the basis b_0 is

$$\begin{pmatrix} 1 & t^{\frac{1}{2}} \\ t^{\frac{1}{2}} & -1 \end{pmatrix}, \; t = p - 1, \tag{16}$$

which we will write as

$$p^{\frac{1}{2}} \, M_0. \tag{17}$$

Finally, the action of F on the 1-dimensional subspace b_r is given by

$$F(x_r) = G(x_r)x_r, \tag{18}$$

where

$$G(x_r) = \begin{cases} p^{\frac{1}{2}}, & p \equiv 1 \, mod \, 4 \\ p^{\frac{1}{2}} i, & p \equiv 3 \, mod \, 4. \end{cases} \tag{19}$$

The constant complex $G(x_r)$ is called the *Legendre symbol mod p* and plays an important role in *Gauss Reciprocity*. The proof of formula (19) can be found in [1]. Set

$$M_r = \begin{cases} I_1, & p \equiv 1 \, mod \, 4 \\ iI_1, & p \equiv 3 \, mod \, 4 \end{cases}. \tag{20}$$

We summarize the discussion in the following theorem.

Theorem 1. The matrix of the Fourier transform F of Z/p relative to the orthonormal basis (6) is the matrix direct sum

$$p^{\frac{1}{2}} \sum_{k=0}^{r} \bigoplus M_k, \tag{21}$$

where

$$M_0 = \begin{pmatrix} (1/p)^{1/2} & ((p-1)/p)^{1/2} \\ ((p-1)/p)^{1/2} & -(1/p)^{1/2} \end{pmatrix}, \tag{22}$$

$$M_k = \begin{pmatrix} 0 & (-1)^k \\ 1 & 0 \end{pmatrix}, \; 1 \le k < r, \tag{23}$$

$$M_r = \begin{cases} I_1, & p \equiv 1 \bmod 4 \\ i\,I_1, & p \equiv i \bmod 4 \end{cases}. \tag{24}$$

We will now diagonalize the matrices M_k, $0 \le k < r$, by orthogonal matrices in the sense given by the formulas below. Set

$$O^+ = \frac{1}{\sqrt{2}} \begin{pmatrix} 1 & 1 \\ 1 & -1 \end{pmatrix}, \tag{25}$$

$$O^- = \frac{1}{\sqrt{2}} \begin{pmatrix} 1 & i \\ i & 1 \end{pmatrix}, \tag{26}$$

O^+ and O^- are orthogonal matrices. Direct computation shows that

$$O^+ M_k (O^+)^{-1} = \begin{pmatrix} 1 & 0 \\ 0 & -1 \end{pmatrix}, \quad k \text{ even}, \tag{27}$$

$$O^- M_k (O^-)^{-1} = \begin{pmatrix} i & 0 \\ 0 & -i \end{pmatrix}, \quad k \text{ odd}. \tag{28}$$

The orthogonal matrix diagonalization of M_0 is more difficult to write. Set

$$a = \left(\frac{1}{p}\right)^{\frac{1}{2}}, \tag{29}$$

$$b = \left(\frac{p-1}{p}\right)^{\frac{1}{2}}. \tag{30}$$

Then we can rewrite M_0 as

$$M_0 = \begin{pmatrix} a & b \\ b & -a \end{pmatrix}. \tag{31}$$

Set

$$O_0 = \frac{1}{\sqrt{2}} \begin{pmatrix} \sqrt{1+a} & \frac{b}{\sqrt{1+a}} \\ \frac{-b}{\sqrt{1+a}} & \sqrt{1+a} \end{pmatrix}. \tag{32}$$

By direct computation

$$O_0 M_0 O_0^{-1} = \begin{pmatrix} 1 & 0 \\ 0 & -1 \end{pmatrix}. \tag{33}$$

Form the matrix direct sum

$$O = \sum_{k=0}^{r} \bigoplus O_k, \tag{34}$$

where

$$O_k = \begin{cases} O^+, \ k \text{ even}, \ 1 \le k < r \\ O^{-1}, \ k \text{ odd} \end{cases} \tag{35}$$

$$O_r = I_1. \tag{36}$$

By the preceeding discussion, O is an orthogonal matrix diagonalizing the matrix (21) given in theorem 1. Applying O to the basis (6) we construct an orthonormal basis diagonalizing the Fourier transform F of Z/p.

14.6. <u>Orthogonal Basis Diagonalizing FT</u> $N = p^m$

Assume $m > 1$. Then

$$L(Z/p^m) = W \ \oplus L_0, \tag{1}$$

where $L_0 = L(1, m - 1)$ is the F-invariant subspace of M-decimated and M^{m-1}-periodic functions on Z/p^m and W is the orthogonal complement. The multiplicative characters and their Fourier transforms will be used to define an orthogonal basis of W diagonalizing the action of F.

Consider the set of non-primitive multiplicative characters of Z/p^m given by the set difference

$$V = \hat{U} \ - \ V(m), \tag{2}$$

and set

$$V' = \{g_x \ : \ x \in V \}. \tag{3}$$

<u>Theorem 1.</u> The set

$$\Omega = V(m) \ \cup \ V \ \cup \ V' \tag{4}$$

is an orthogonal basis of the subspace W.

Proof In section 1, we showed that $\hat{U}$ is an orthogonal set. Since $p^{-\frac{m}{2}} F$ preserves inner products, the formulas of section 4, imply that the set V' is orthogonal. The vanishing of the functions in V' on the unit group U implies that V' is orthogonal to $\hat{U}$ in the sense that

$$< \hat{U}, V' > = 0.$$

The same argument shows that $L(1, m-1)$ is orthogonal to $\hat{U}$. Since $L(1, m-1)$ is F-invariant, $L(1, m-1)$ is orthognal to $F(\hat{U})$ proving that Ω is orthogonal to $L(1, m-1)$ and

$$\Omega = W.$$

Dimension arguments show that Ω is a basis of W. First

$$dim(W) = p^m - p^{m-2} = p^{m-2}(p^2 - 1).$$

Since, we also have

$$o(\Omega) = o(\hat{U}) + o(V') = p^{m-1}(p-1) + p^{m-2}(p-1) = p^{m-2}(p^2 - 1),$$

the theorem follows.

An orthonormal basis will be constructed by scalar multiplication of the orthogonal basis Ω. Fix a generator z of the unit group U of Z/p^m. Then

$$V(m) = \{x_k : 0 \le k < t, \ (p, k) = 1\}, \tag{5}$$

where $t = o(U)$. Set

$$B(m) = \{x_k : 0 \le k < r, \ (p, k) = 1\}. \tag{6}$$

Since $V(m)$ has no real multiplicative characters, we have the disjoint union

$$V(m) = B(m) \cup B^*(m). \tag{7}$$

Theorem 1 and the formulas of the preceeding section imply that the disjoint union

$$\{x \in B(m) \cup V\} \bigcup \{F(x) : x \in B(m) \cup V\} \tag{8}$$

is an orthogonal basis of the subspace W. Since

$$< x, x > = t, \tag{9}$$

$$< F(x), F(x) > = p^m t, \tag{10}$$

the set

$$b(x) = \{t^{-\frac{1}{2}} x, t^{-\frac{1}{2}} p^{-\frac{m}{2}} F(x)\} \tag{11}$$

is orthonormal and the set

$$\{b(x) : x \in B(m) \cup V\} \tag{12}$$

is an orthonormal basis of the subspace W. The action of the Fourier transform F will now be described relative to the basis (12).

In general, for any $f \in L(Z/p^m)$,

$$F^2(f)(a) = p^m f(-a), \ a \in Z/p^m. \tag{13}$$

Applying the formula to multiplicative characters, we have

$$F^2(x) = p^m x(-1)x, \ x \in \hat{U}. \tag{14}$$

Then the space

$$W(x) = Ln(b(x)) \tag{15}$$

is F-invariant and the matrix of the action of F relative to the basis $b(x)$ is

$$p^{\frac{m}{2}} M(x), \tag{16}$$

where

$$M(x) = \begin{pmatrix} 0 & x(-1) \\ 1 & 0 \end{pmatrix}. \tag{17}$$

Theorem 2. The matrix of the Fourier transform F of Z/p^m, $m > 1$, relative to the orthonormal basis (12) is the matrix direct sum

$$p^{\frac{m}{2}} \sum_x \oplus M(x), \tag{18}$$

where x runs over $B(m) \cup V$.

Since $x(-1) = \pm 1$, the matrix $M(x)$ is either

$$M^+ = \begin{pmatrix} 0 & 1 \\ 1 & 0 \end{pmatrix}, \tag{19}$$

$$M^- = \begin{pmatrix} 0 & -1 \\ 1 & 0 \end{pmatrix}. \tag{20}$$

The first is diagonalized by the orthogonal matrix

$$O^+ = \frac{1}{\sqrt{2}} \begin{pmatrix} 1 & 1 \\ 1 & -1 \end{pmatrix}, \tag{21}$$

and the second is diagonalized by the matrix

$$O^- = \frac{1}{\sqrt{2}} \begin{pmatrix} 1 & 1 \\ i & -i \end{pmatrix}. \tag{22}$$

Explicitly,

$$(O^+)^{-1} M^+ O^+ = \begin{pmatrix} 1 & 0 \\ 0 & -1 \end{pmatrix}, \tag{23}$$

$$(O^-)^{-1} M^- O^- = \begin{pmatrix} -i & 0 \\ 0 & i \end{pmatrix}. \tag{24}$$

Setting

$$O = \sum_x \oplus O(x), \tag{25}$$

where

$$O(x) = \begin{cases} O_1, & x(-1) = 1 \\ O_2, & x(-1) = -1 \end{cases}, \tag{26}$$

we have that the matrix O diagonalizes the matrix (18). Applying O to the basis (12) determines an orthonormal basis of the subspace

W diagonalizing the restriction of F to W. The details are left to the problems.

Since the restruction of F restricted to $L(1, m-1)$, relative to the orthonormal basis is $F(p^{m-2})$. We can use induction to find an orthonormal basis of $L(Z/p^m)$ diagonalizing F.

[References]

[1] Tolimieri, R. "Multiplicative Characters and the Discrete Fourier Transform", *Adv. in Appl. Math.* 7 (1986), 344-380.

[2] Auslander, L. Feig, E. and Winograd, S. "The Multiplicative Complexity of the Discrete Fourier Transform", *Adv. in Appl. Math.* 5 (1984), 31-55.

[3] Rader, C. "Discrete Fourier Transforms When the Number of Data Samples is Prime", *Proc. of IEEE*, 56 (1968), 1107-1108.

[4] Winograd, S. *Arithmetic Complexity of Computations*, CBMS Regional Conf. Ser. in Math. Vol. 33, Soc. Indus. Appl. Math., Philadelphia, 1980.

[5] Tolimieri, R. "The Construction of Orthogonal Basis Diagonalizing the Discrete Fourier Transform", *Adv. in Appl. Math.* 5 (1984), 56-86.

Problems

1. Find a generator z for unit group U of $Z/11$, order U exponentially relative to z.

2. Define the 10 multiplicative characters of $Z/11$.

3. Find a generator z for unit group U of $Z/13$, order U exponentially relative to z.

4. Define the 12 multiplicative characters of $Z/13$.

5. Find a generator z for unit group U of $Z/3^2$, order U exponentially relative to z. Define the 6 multiplicative characters of $Z/3^2$.

6. Find a generator z for unit group U of $Z/5^2$, order U exponentially relative to z. Define the 20 multiplicative characters of $Z/5^2$.

7. Give the Gauss sums $G_k = G_1(x_k)$, $k = 1, 2, 3, 4, 5$ for $p = 7$.

Chapter 15

RATIONALITY

Multiplicative character theory provides a natural setting for developing the complexity results of Auslander-Feig-Winograd [1]. The first reason for this is the simplicity of the formulas describing the action of FT on multiplicative characters. We will now discuss a second important property of multiplicative characters. In a sense defined below, the spaces spanned by the subsets V_k, $1 \leq k < m$, of multiplicative characters are rational subspaces. As a consequence, we will be able to rationally manipulate the FT matrix $F(p^m)$ into block diagonal matrices where each block action corresponds to some polynomial multiplication modulo a rational polynomial of a special kind. This is the main result in the work of Auslander-Feig-Winograd. Details from the point of view of multiplicative character theory can be found in [2].

Although these results proceed in a straight forward fashion, notation becomes complicated. After some preliminary general definitions we will derive in detail some examples. A function $f \in L(Z/N)$ is called a *rational vector* if the standard representation of f is an N-tuple of rational numbers. A subspace X of $L(Z/N)$ is called a *rational subspace* if X has a basis consisting solely of rational vectors. Such a basis is called a *rational basis* of X.

15.1. An Example $N = 7$

Take $p = 7$ and $z = 3$ as a generator of $U = U(7)$. Let $V' = \widehat{U} - \{x_0\}$. Then V' is the set of 5 multipicative characters given by the following table,

	0	1	3	2	6	4	5
x_1	0	1	v	v^2	v^3	v^4	v^5
x_2	0	1	v^2	v^4	v^6	v^8	v^{10}
x_3	0	1	v^3	v^6	v^9	v^{12}	v^{15}
x_4	0	1	v^4	v^8	v^{12}	v^{16}	v^{20}
x_5	0	1	v^5	v^{10}	v^{15}	v^{20}	v^{25}

$v = e^{(2\pi i/6)}$.

A rational basis $R(1)$ will be constructed for the linear span $Ln(V')$. Define the functions

$$R(1) = \{r_k : 0 \le k < 5\}, \tag{1}$$

by

$$r_j = e_{3j} - e_5. \tag{2}$$

These functions are given by the following table,

	0	1	3	2	6	4	5
r_0	0	1	0	0	0	0	-1
r_1	0	0	1	0	0	0	-1
r_2	0	0	0	1	0	0	-1
r_3	0	0	0	0	1	0	-1
r_4	0	0	0	0	0	1	-1

At the point 0, each of these functions takes on the value 0.

We will now show that for each $x \in V(1)$,

$$x = \sum_{j=0}^{4} x(3^j) r_j. \tag{3}$$

By definition, the formula holds at the points

$$1, 3, 2, 6, 4. \tag{4}$$

At the point $5 \equiv 3^5$ we have

$$\sum_{j=0}^{4} x(3^j) r_j(5) = - \sum_{j=0}^{4} x(3^j). \tag{5}$$

Since $x \neq x_0$,

$$\sum_{j=0}^{5} x(3^j) = 0, \tag{6}$$

implying that (3) holds at the point 5.

Using formula (4), we have

$$\begin{bmatrix} x_1 \\ x_2 \\ x_3 \\ x_4 \\ x_5 \end{bmatrix} = X(1) \begin{bmatrix} r_0 \\ r_1 \\ r_2 \\ r_3 \\ r_4 \end{bmatrix}, \tag{7}$$

where since $x_k(3^j) = v^{kj}$,

$$X(1) = \begin{bmatrix} 1 & v & v^2 & v^3 & v^4 \\ 1 & v^2 & v^4 & v^6 & v^8 \\ 1 & v^3 & v^6 & v^9 & v^{12} \\ 1 & v^4 & v^8 & v^{12} & v^{16} \\ 1 & v^5 & v^{10} & v^{15} & v^{20} \end{bmatrix}. \tag{8}$$

We will also write (7) as

$$V(1) = X(1)R(1), \tag{9}$$

the matrix $X(1)$ is a *Van der Monde matrix* and is non-singular. It follows that $R(1)$ is a basis of $Ln(V(1))$ and $X(1)$ is the change of basis matrix.

Consider the restriction of the FT F of $Z/7$ to the subspace $Ln(V(1))$. By theorem 14.3.2.

$$F(x_j) = G_1(x_j)x_j^*, \quad 1 \leq j < 6, \tag{10}$$

implying that the matrix of F restricted to $Ln(V(1))$ relative to the basis $V(1)$ is the *skew-diagonal matrix*

$$G(1) = \begin{bmatrix} & & & & G_1(x_5) \\ & & & 0 & \bullet \\ & & \bullet & & \\ & G_1(x_2) & & 0 & \\ G_1(x_1) & & & & \end{bmatrix}. \tag{11}$$

By (10), the matrix of F restricted to $Ln(V(1))$ relative to the basis $R(1)$ is

$$Y(1)G(1)X(1)^{-1}. \tag{12}$$

Complete $R(1)$ to the rational basis

$$e_0, \ e_0 + x_0; \ R(1), \tag{13}$$

of $L(Z/7)$. The matrix of F relative to this rational basis is the matrix direct sum

$$\begin{pmatrix} 0 & 7 \\ 1 & 0 \end{pmatrix} \oplus [X(1)G(1)X(1)^{-1}]. \tag{14}$$

Two matrices A and B are called *rationally related*, if

$$A = Q_1 B Q_2, \tag{15}$$

where Q_1 and Q_2 are rational non-singular matrices. In classical complexity theory, rational multiplications are free and rationally related matrices have the same multiplicative complexity. In particular, $F(7)$ is rationally related to (14) and the multiplicative complexity of $F(7)$ is equal to the multiplicative complexity of

$$Y(1) \equiv X(1)G(1)X(1)^{-1}. \tag{16}$$

15.2. Prime Case

The methods used in the example extend in a straightfoward manner to any prime p. Take a generator z of the unit group U of Z/p. Consider the set of primitive multiplicative characters

$$V(1) = \{ x_k \ : \ 1 \le k < t \}, \ t = p - 1. \tag{1}$$

A rational basis

$$R(1) = \{ r_k \ : \ 0 \le k < t - 1 \}, \tag{2}$$

for $Ln(V(1))$ can be constructed by setting

$$r_k = e_{z^k} - e_{z^{t-1}}, \ 0 \le k < t - 1. \tag{3}$$

In fact, arguing as in example, we have, for any $x \in V(1)$,

$$x = \sum_{k=0}^{t-2} x(z^j) r_k. \tag{4}$$

Applying this formula with $x = x_k$, $1 \le k < t$, we have the matrix formula

$$V(1) = X(1) \, R(1), \tag{5}$$

where $v = exp(2\pi i/t)$ and

$$X(1) = \begin{bmatrix} 1 & v & \cdots & v^{t-2} \\ 1 & v^2 & \cdots & v^{2(t-1)} \\ \cdots & & & \\ 1 & v^{t-1} & \cdots & v^{(t-1)(t-2)} \end{bmatrix}. \tag{6}$$

The Van der Monde matrix $X(1)$ is non-singular and $R(1)$ is a rational basis of $Ln(V(1))$.

The matrix of the restriction of the FT F to the subspace $LN(V(1))$ relative to the basis $V(1)$ is the skew-diagonal matrix

$$\begin{bmatrix} & & & G_1(x_{t-1}) \\ & 0 & \bullet & \\ & \bullet & & \\ & \bullet & 0 & \\ G_1(x_1) & & & \end{bmatrix}. \tag{7}$$

We are lead to the following theorem.

Theorem 1. The matrix of the FT F of Z/p relative to the rational basis of $L(Z/p)$

$$e_0, e_0 + x_0; \ R(1), \tag{8}$$

is the matrix direct sum

$$\begin{bmatrix} 0 & p \\ 1 & 0 \end{bmatrix} \oplus [X(1) \, G(1) \, X(1)^{-1}]. \tag{9}$$

It follows that $F(p)$ is rationally related to the matrix (9) and has multiplicative complexity equal to the multiplicative complexity of

$$Y(1) = X(1)G(1)X(1)^{-1}. \tag{10}$$

The relationship between $Y(1)$ and polynomial multiplication algorithms can be found in [2].

A linear isomorphism P of $L(Z/n)$ is called a permutation, if relative to the standard basis P is a permutation matrix. The skew-circulant matrix $G(1)$ can be replaced by a diagonal matrix by introducing a permutation transformation P defined by the formulas

$$P(e_0) = e_0, \tag{11}$$

$$P(e_{z^k}) = e_{z^{-k}}, \ 0 \le k < t = p - 1, \tag{12}$$

then

$$P(x_k) = x_k^*, \ 0 \le k < t. \tag{13}$$

Consider the matrix PF. The action of PF restricted to the subspace $Ln(e_0, e_0 + x_0)$ is the same as the action of F restricted to this subspace. The action of PF on the basis $V(1)$ of $Ln(V(1))$ is given by the diagonal matrix

$$D(1) = \begin{bmatrix} G_1(x_1) & & & & \\ & G_1(x_2) & & 0 & \\ & & \ddots & & \\ & 0 & & & \\ & & & & G_1(x_{t-1}) \end{bmatrix}, \tag{14}$$

proving the next theorem.

Theorem 2. The matrix of PF relative to the rational basis

$$e_0, \ e_0 + x_0 \ ; \ R(1) \tag{15}$$

is the matrix direct sum

$$\begin{bmatrix} 0 & p \\ 1 & 0 \end{bmatrix} \oplus [X(1) \, D(1) \, X(1)^{-1}]. \tag{16}$$

15.3. <u>An Example</u> $N = 3^2$

Take $z = 2$ as a generator of the unit group U of $Z/9$ and consider the set of primitive multiplicative characters

$$V(2) = \{x_1, x_2, x_4, x_5\}. \tag{1}$$

We repeat the defining table,

	0	3	6	1	2	4	8	7	5
x_1	0	0	0	1	v	v^2	v^3	v^4	v^5
x_2	0	0	0	1	v^2	v^4	v^6	v^8	v^{10}
x_4	0	0	0	1	v^4	v^8	v^{12}	v^{16}	v^{20}
x_5	0	0	0	1	v^5	v^{10}	v^{15}	v^{20}	v^{25}

where $v = exp(2\pi i/6)$.

Define the set of functions

$$R(2) = \{r_0, r_1, r_2, r_3\}, \tag{2}$$

by the following table,

	0	3	6	1	2	4	8	7	5
r_0	0	0	0	1	0	0	0	-1	0
r_1	0	0	0	0	1	0	0	0	-1
r_2	0	0	0	0	0	1	0	-1	0
r_3	0	0	0	0	0	0	1	0	-1

The functions are also defined by the formulas

$$r_0 = e_1 - e_7, \tag{3}$$

$$r_1 = e_2 - e_5, \tag{4}$$

$$r_2 = e_4 - e_7, \tag{5}$$

$$r_3 = e_8 - e_5. \tag{6}$$

We claim

$$x_k = \sum_{l=0}^{3} v^{lk} r_l, \; k = 1, 2, 4, 5. \tag{7}$$

By definition, (7) holds at the points

$$0, \, 3, \, 6, \, 1, \, 2, \, 4, \, 8. \tag{8}$$

At the point $7 \equiv 2^4$,

$$x_k(7) = v^{4k}, \tag{9}$$

$$\sum_{l=0}^{3} v^{lk} r_l(7) = -1 - v^{2k}, \; k = 1, 12, 4, 5. \tag{10}$$

Since $v \neq 1$,

$$1 + v^{2k} + v^{4k} = 0, \; k = 1, 2, 4, 5, \tag{11}$$

implying (7) holds at the point 7. The same argument shows that (7) holds at the point 5, completing the proof of formula (7).

Set

$$X(2) = \begin{bmatrix} 1 & v & v^2 & v^3 \\ 1 & v^2 & v^4 & v^6 \\ 1 & v^4 & v^8 & v^{12} \\ 1 & v^5 & v^{10} & v^{15} \end{bmatrix}, \; v = exp(2\pi i/6). \tag{12}$$

Then

$$\begin{bmatrix} x_1 \\ x_2 \\ x_4 \\ x_5 \end{bmatrix} = X(2) \begin{bmatrix} r_0 \\ r_1 \\ r_2 \\ r_3 \end{bmatrix}, \tag{13}$$

which we can rewrite as

$$V(2) = X(2)\, R(2). \tag{14}$$

Since $X(2)$ is non-singular, $R(2)$ is a rational basis of $Ln(V(2))$.

We complete $R(2)$ to a rational basis of $L(Z/9)$

$$f \; ; \; x_0, g_0 \; ; \; x_3, g_3 \; ; \; R(2), \tag{15}$$

where

$$f = e_0 + e_3 + e_6, \qquad (16)$$

and $g_k = g_{x_k}$, $k = 0, 3$.

	0	3	6	1	2	4	8	7	5
f	1	1	1	0	0	0	0	0	0
x_0	0	0	0	1	1	1	1	1	1
g_0	−2	1	1	0	0	0	0	0	0
x_3	0	0	0	1	−1	1	−1	1	−1
g_3	0	1	−1	0	0	0	0	0	0

We will now compute the matrix of the FT F of $Z/9$ relative to the rational basis (15). The formulas of section 4, chapter 14, can be used to compute the action of F on multiplicative characters which can then be used to compute the action of F on the rational basis applying the change of basis matrix $X(2)$.

First consider the action of F on $Ln(V(2))$, the space spanned by the primitive multiplicative characters. From chapter 14, we have the formulas

$$F(x_k) = G_2(x_k)x_k^*, \ k = 1, 2, 4, 5. \qquad (17)$$

This implies that the matrix of the restriction of F to $Ln(V(2))$ relative to the basis $V(2)$ is the skew-diagonal

$$G(2) = \begin{bmatrix} & & & 0 & & G_2(x_5) \\ & & & & G_2(x_4) & \\ & & G_2(x_2) & & & 0 \\ G_2(x_1) & & & & & \end{bmatrix}. \qquad (18)$$

Relative to the basis $R(2)$, the corresponding matrix is

$$X(2)\, G(2)\, X(2)^{-1}. \qquad (19)$$

Applying formulas from section 3, chapter 14,

$$F(x_0) = -3g_0, \qquad (20)$$

$$F(x_3) = 3\sqrt{3}\, i\, g_3. \tag{21}$$

Since

$$F^2(h)(a) = 9h(-a),\ a \in Z/9,\ h \in L(Z/9), \tag{22}$$

we have

$$F(g_0) = -3x_0, \tag{23}$$

$$F(g_3) = \sqrt{3}ix_3.$$

Finally, by direct computation

$$F(f) = 3f. \tag{24}$$

Putting together formulas $(19-24)$, we have that the matrix of F relative to the rational basis (15) is the matrix direct product sum

$$[3] \oplus (-3) \begin{bmatrix} 0 & 1 \\ 1 & 0 \end{bmatrix} \oplus \sqrt{3}i \begin{bmatrix} 0 & 1 \\ 3 & 0 \end{bmatrix} \oplus [X(2)G(2)X(2)^{-1}]. \tag{25}$$

15.4. Transform Size $N = p^2$

Fix an odd prime p and take z as a generator of the unit group U of Z/p^2. Consider the set of primitive multiplicative characters

$$V(2) = \{\, x_k : 0 \le k < t,\ (p, k) = 1 \,\}, \tag{1}$$

where $t = p(p-1)$. The dimension of $Ln(V(2))$ is

$$s = (p-1)^2. \tag{2}$$

A rational basis for $Ln(V(2))$

$$R(2) = \{ r_k : 0 \le k < s \}, \tag{3}$$

is defined by the following table:

	1	z	$\dots$	z^{s-1}	z^s	$\dots$	z^{t-1}
r_0							
r_1			I_s		$-1_{p-1} \otimes I_{p-1}$		
$\vdots$							
$\vdots$							
r_{s-1}							

These functions vanish off of U. Arguing as in the preceeding example, we have the formula

$$x_k = \sum_{l=0}^{s-1} v^{lk} r_l, \ 0 \le k < t, \ (p, k) = 1, \tag{4}$$

where $v = exp(2\pi i/t)$. We can rewrite (9) in matrix form as

$$V(2) = X(2)R(2), \tag{5}$$

where

$$X(2) = |v^{lk}|_{0 \le k < t, \, 0 \le l < s, \, (p,k)=1}. \tag{6}$$

The non-singularity of the Van der Monde matrix $X(2)$ implies that $R(2)$ is a basis of $Ln(V(2))$.

Consider the action of the FT F of Z/p^2. By chapter 14,

$$F(x_k) = G_2(x_k)x_k^*, \ x_k \in V(2). \tag{7}$$

It follows that the matrix of the restriction of F to $Ln(V(2))$ relative to the basis $V(2)$ is the skew-diagonal matrix

$$G(2) = \begin{bmatrix} & & & & G_2(x_{t-1}) \\ & & & \ddots & \\ & & G_2(x_j) & & \\ & \ddots & & & \\ G_2(x_1) & & & & \end{bmatrix}_{0 \le j < t, \, (p,k)=1}. \tag{8}$$

Applying the change of basis formula (5), the matrix of the restriction to $Ln(V(2))$ relative to the basis $R(2)$ is

$$X(2)\, G(2)\, X(2)^{-1}. \tag{9}$$

The unique ideal of Z/p^2 is

$$M = p\, Z/p^2. \tag{10}$$

Consider the set

$$V_1 = \{\, x_{pk} : 0 \le k < p - 1 \,\}, \tag{11}$$

and the set

$$V' = \{ g_{pj} : 0 \le k < p - 1 \,\}, \tag{12}$$

where

$$g_{pk}(pa) = x_{pk}(a),\ a \in Z/p^2. \tag{13}$$

Up to constant multiples, given by the formulas of section 4, chapter 14, the set V' consists of FT's of the set V. In fact, we have

$$F(x_{pk}) = p\, G_1(x_{pk}) g_{pk}^*, \tag{14}$$

by theorem 2, section 6 of chapter 14.

The periodization

$$\pi_1^* : L(Z/p) \mapsto L(Z/p^2), \tag{15}$$

defined in chapter 13, bijectively maps the set $V(1)$ of primitive multiplicative characters of Z/p onto V_1 and induces to a linear isomorphism of $Ln(V(1))$ onto $Ln(V_1)$. Since by section 2, the set $R(1)$ is a rational basis of $Ln(V(1))$, the periodization

$$R_1 = \pi_1^*(R(1)), \tag{16}$$

is a rational basis of $Ln(V_1)$. Explicitly

$$R_1 = \{ r_k^{(1)} : 0 \le k < p - 2 \,\}, \tag{17}$$

where each function in R_1 is M-periodic and has values given by the following table,

	1	z	$\cdots$	$\cdots$	z^{p-3}	z^{p-2}
$r_0^{(1)}$	1	0	$\cdots$	$\cdots$	0	-1
$r_1^{(1)}$	0	1	0	$\cdots$	0	-1
$\vdots$			$\ddots$			
$r_{p-3}^{(1)}$	0	$\cdots$	$\cdots$	0	1	-1

By (5) of section 2 ,

$$V_1 = X(1)\, R_1, \tag{18}$$

and R_1 is a rational basis on $Ln(V_1)$.

Arguing in the same way, V_1' is bijectively equivalent to $V(1)$ under the decimation

$$\sigma_1^* \; : \; L(Z/p) \longmapsto L(Z/p^2), \tag{19}$$

and a rational basis

$$S_1 = \left\{ s_k^{(1)} \, : \, 0 < k < p - 2 \right\}, \tag{20}$$

for $Ln(V_1')$ is defined by the following table,

	p	pz	$\cdots$	$\cdots$	pz^{p-3}	pz^{p-2}
$s_0^{(1)}$	1	0	$\cdots$	$\cdots$	0	-1
$s_1^{(1)}$	0	1	0	$\cdots$	0	-1
$\vdots$			$\ddots$			$\vdots$
$s_{p-3}^{(1)}$	0	$\cdots$	$\cdots$	0	1	-1

These functions vanish off of the set pU. Analogous to (18), we have

$$V_1' = X(1)S_1. \tag{21}$$

Formula (11) contains all the information needed to determine the matrix of F restricted to $Ln(V_1 \cup V_1')$ relative to the basis $V_1 \cup V_1'$.

The change of basis formulas (18) and (20) will then be applied to find the matrix of this restriction relative to the rational basis $R_1 \cup S_1$.

Applying the formula

$$(F^2 f)(a) = p^2 f(-a), a \in Z/p^2, f \in L(Z/p^2), \qquad (22)$$

to (14) we have

$$F(g_{pk}) = G_1(x_{pk})x_{pk}^*. \qquad (23)$$

From (14) and (22), the matrix of the restriction of F to the subspace $Ln(V_1 \cup V_1')$ relative to the basis $V_1 \cup V_1'$ is the skew-diagonal matrix

$$G_1 = \begin{bmatrix} 0 & G(1) \\ p\,G(1) & 0 \end{bmatrix}. \qquad (24)$$

Relative to the rational basis

$$R_1, \ S_1, \qquad (25)$$

the matrix of F is

$$\big(X(1) \oplus X(1)\big)G_1\big(\,X(1)^{-1}\big) \oplus X(1)^{-1}). \qquad (26)$$

From theorem 4.3 of chapter 14,

$$F(x_0) = -pg_0, \qquad (27)$$

where g_0 is the rational vector defined at that time. Applying (21)

$$F(g_0) = -p\,x_0. \qquad (28)$$

It follows that the matrix of the restriction of F to the subspace $Ln(x_0, g_0)$ relative to the basis

$$\{x_0, g_0\}, \qquad (29)$$

is the matrix

$$-p \begin{pmatrix} 0 & 1 \\ 1 & 0 \end{pmatrix}. \qquad (30)$$

Finally, the rational vector

$$f = \sum_{k \in M} e_k, \tag{31}$$

spans the 1-dimensional subspace $L(1,1)$ and direct computation shows that

$$F(f) = pf. \tag{32}$$

The following result has been proved.

Theorem 1. The matrix of the Fourier transform F of Z/p^2 relative to the rational basis

$$f \; ; x_0, g_0 \; ; R_1, S_1 \; ; R(2), \tag{33}$$

is the matrix direct sum

$$[p] \oplus (-p) \begin{pmatrix} 0 & 1 \\ 1 & 0 \end{pmatrix} \oplus [X^{(2)}(1)G_1 X^{(2)}(1)^{-1}]$$

$$\oplus [X(2)G(2)X(2)^{-1}], \tag{34}$$

where

$$G_1 = \begin{bmatrix} 0 & G(1) \\ pG(1) & 0 \end{bmatrix},$$

$$X^{(2)}(1) = X(1) \oplus X(1). \tag{35}$$

Define the permutation transformation P by the formulas

$$P(e_0) = e_0, \tag{36}$$

$$P(e_{z^k}) = e_{z^{-k}}, \; 0 \le k < p(p-1), \tag{37}$$

$$P(e_{pz^k}) = e_{pz^{-k}}, \; 0 \le k < p - 1. \tag{38}$$

Direct computations shows that

$$P(x_k) = x_k^*, \; 0 \le k < p(p-1), \tag{39}$$

$$P(g_{pk}) = g^*_{pk}, \ 0 \le k < p - 1. \qquad (40)$$

Then

$$PF(x_k) = G_2(x_k)x_k, \ 0 \le k < p(p-1), \ (p,k) = 1, \qquad (41)$$

$$PF(x_{pk}) = pG_1(x_{pk})\,g_{pk}, \ 0 < k < p - 1, \qquad (42)$$

$$PF(g_{pk}) = G_1(x_{pk})x_{pk}, \ 0 < k < p - 1. \qquad (43)$$

Since, we also have

$$P(x_0) = x_0, \qquad (44)$$

$$P(g_0) = g_0, \qquad (45)$$

$$P(f) = f, \qquad (46)$$

we have the next result.

Theorem 2. The matrix of PF relative to the rational basis (33) is the matrix direct sum

$$|p| \oplus (-p) \begin{pmatrix} 0 & 1 \\ 1 & 0 \end{pmatrix} \oplus [X^{(2)}(1)\, D_1\, X^{(2)}(1)^{-1}]$$

$$\oplus [X(2)D(2)X(2)^{-1}]. \qquad (47)$$

where

$$D(2) = \begin{bmatrix} G_2(x_1) & & & & \\ & \ddots & & & \\ & & G_2(x_2) & & \\ & & & \ddots & \\ & & & & G_2(x_{t-1}) \end{bmatrix}_{0 \le k < t, \ (p,k)=1}$$
$$(48)$$

$$D_1 = \begin{bmatrix} 0 & D(1) \\ pD(1) & 0 \end{bmatrix}. \qquad (49)$$

Summarizing, the multiplicative complexity of $F(p^2)$ is the multiplicative complexity of

$$[X^{(2)}(1)\, D_1\, X^{(2)}(1)^{-1}] \oplus X(2)D(2)X(2)^{-1}. \qquad (50)$$

15.5. Exponential Basis

The results of this chapter can be referred to the exponential basis. First take z as a generator of the unit group U of Z/p and set

$$E^{\times}(1) = \{e_{z^k} : 0 \le k < p-1\},\tag{1}$$

$$L(1) = [I_{p-2}| - 1_{p-2}].\tag{2}$$

Then formula (3) of section 2 can be rewritten in matrix form as

$$R(1) = L(1)E^{\times}(1).\tag{3}$$

More generally, if z is a generator of the unit group U of Z/p^2 and

$$E^{\times}(2) = \{e_{z^k} : 0 \le k < p(p-1)\}.\tag{4}$$

Set

$$L(2) = |I_s| - 1_{p-1} \otimes I_{p-1}|, \quad s = (p-1)^2.\tag{5}$$

Then by the table in section 4, defining $R(2)$,

$$R(2) = L(2)E^{\times}(2).\tag{6}$$

The table defining R_1 in section 4 can be rewritten in matrix form as

$$R_1 = \left(1_p^t \otimes L(1)\right)E^{\times}(2).\tag{7}$$

15.6. Polynomial Product modulo a Polynomial

We will show how the action of the matrices

$$X(1)D(1)X(1)^{-1},\tag{1}$$

$$X(2)D(2)X(2)^{-1},\tag{2}$$

can be carried out by polynomial product modulo special integer polynomials. Let

$$v_0, \ v_1, \ \cdots, \ v_{n-1}, \tag{3}$$

be distinct complex numbers and form the polynomial

$$g(u) = (u - v_0)(u - v_1) \cdots (u - v_{n-1}). \tag{4}$$

As a complex vector space, the quotient polynomial ring $\mathbb{C}[u]/g(u)$ has dimension n and the set

$$\{u^l : \quad 0 \le l < n\}, \tag{5}$$

is a basis. As a special case of the CRT, we have the next result.

CRT. There exists a basis

$$\{E_k : 0 \le k < n\}, \tag{6}$$

of $\mathbb{C}[u]/g(u)$ satisfying

$$E_j(v_l) = \begin{cases} 1, & l=k, \\ 0, & l \ne; \end{cases} \quad 0 \le l, k < n. \tag{7}$$

From (7), we have

$$E_k E_l = \begin{cases} 1, & l=k, \\ 0, & l \ne k; \end{cases} \quad 0 \le l, k < n. \tag{8}$$

The basis (5) and (6) are related by

$$u^l = \sum_{k=0}^{n-1} v_k^l E_k, \quad 0 \le l < n. \tag{9}$$

Define the Van Der Monde matrix

$$Z = \begin{bmatrix} 1 & 1 & \cdots & 1 \\ v_0 & v_1 & \cdots & v_{n-1} \\ \vdots & & & \\ v_0^{n-1} & v_1^{n-1} & \cdots & v_{n-1}^{n-1} \end{bmatrix}, \tag{10}$$

and observe that

$$\begin{bmatrix} 1 \\ u \\ \vdots \\ u^{n-1} \end{bmatrix} = Z \begin{bmatrix} E_0 \\ E_1 \\ \vdots \\ E_{n-1} \end{bmatrix}. \tag{11}$$

Since Z is based on distinct complex numbers, it is a non-singular matrix.

Take $\varphi \in \mathbb{C}\,[u]/g(u)$, for any $\psi \in \mathbb{C}\,[u]/g(u)$, define

$$\tau(\varphi)\psi = \varphi\psi, \tag{12}$$

and observe that $\tau(\varphi)$ is a linear transfromation of $\mathbb{C}\,[u]/g(u)$. Denote the matrix of $\tau(\varphi)$ relative to the basis (5) by $\Re(\varphi)$ and relative to the basis (6) by $\Re'(\varphi)$. Then

$$\Re(\varphi) = (Z^t)^{-1}\Re'(\varphi)Z^t, \tag{13}$$

and after taking transpose

$$\Re^t(\varphi) = Z\Re'(\varphi)Z^{-1}. \tag{14}$$

Since by (7)

$$\varphi = \sum_{k=0}^{n-1} \varphi(v_k)E_k, \tag{15}$$

we have

$$\Re'(\varphi) = diag.[\varphi(v_k)]_{0 \le k < n}. \tag{16}$$

The form of (14) resembles the form of the matrices (1) and (2) in that Z is a Van der Monde matrix and $\Re'(\varphi)$ is a diagonal matrix. To see the exact relationship, we reason as follows. Fix an odd prime p throughout this discussion. Set $v = exp(2\pi i/p - 1)$. Then

$$u^{p-1} - 1 = \prod_{k=0}^{p-2} (u - v^k) = (u - 1)g_1(u), \tag{17}$$

where

$$g_1(u) = \prod_{k=1}^{p-2}(u - v^k). \tag{18}$$

Factorize $(u - 1)$ from $u^{p-1} - 1$, we have

$$g_1(u) = u^{p-2} + u^{p-3} + \cdots + u + 1. \tag{19}$$

Take $g(u) = g_1(u)$ in the discussion above and denote $\Re(\varphi)$ by $\Re_1(\varphi)$ and $\Re'(\varphi)$ by $\Re'_1(\varphi)$. Then

$$\Re_1^t(\varphi) = X(1)\Re'_1(\varphi)X(1)^{-1}, \tag{20}$$

where $X(1)$ is the Van der Monde matrix based on

$$v, v^2, \cdots, v^{p-2}. \tag{21}$$

Define the "Gauss Polynomial"

$$\varphi_1(u) = \sum_{k=0}^{p-2} w^{z^k} u^k, \quad w = exp(2\pi i/p), \tag{22}$$

and observe that

$$\varphi_1(v^k) = G_1(x_k), \quad 0 \le k < p - 1. \tag{23}$$

Then

$$\Re'(\varphi_1) = D(1), \tag{24}$$

and (20) becomes

$$\Re_1^t(\varphi) = X(1)D(1)X(1)^{-1}. \tag{25}$$

In words: the transpose of $X(1)D(1)X(1)^{-1}$ is equivalent to multiplication by $\varphi_1 \bmod g_1(u)$.

Now consider the case $n = p^2$ and take $t = p(p - 1)$. Set $v = exp(2\pi i/t)$. Then

$$u^t - 1 = \prod_{k=0}^{t-1}(u - v^k) = g_2(u) \prod_{0 \le k < t;\, p/k}(u - v^k), \tag{26}$$

where

$$g_2(u) = \prod_{0 \le k < t} (u - v^k), \quad (p, k) = 1. \tag{27}$$

Set $\alpha = v^p = exp(2\pi i/p - 1)$. Then

$$u^{p-1} - 1 = \prod_{k=0}^{p-2} (u - \alpha^k) = \prod_{0 \le k < t;\ p/k} (u - v^k), \tag{28}$$

and we can rewrite (26) as

$$u^t - 1 = g_2(u)(u^{p-1} - 1). \tag{29}$$

Dividing $u^{p-1} - 1$ into $u^t - 1$, we have

$$g_2(u) = u^{(p-1)^2} + u^{(p-2)(p-1)} + \cdots + u^{p-1} + 1. \tag{30}$$

Taking $g(u) = g_2(u)$ in the above discussion, we denote $\Re(\varphi)$ by $\Re_2(\varphi)$ and $\Re'(\varphi)$ by $\Re'_2(\varphi)$, (14) becomes

$$\Re_2^t(\varphi) = X(2)\Re'_2(\varphi)X(2)^{-1}, \tag{31}$$

where $X(2)$ is the Van Der Monde matrix based on

$$v^k, \quad 0 \le k < t, \quad (p, k) = 1. \tag{32}$$

Define the "Gauss Polynomial"

$$\varphi_2(u) = \sum_{j=0}^{t-1} w^{z^k} u^k, \quad w = exp(2\pi i/p^2). \tag{33}$$

Direct computation shows that

$$\Re'_2(\varphi_2) = D(2), \tag{34}$$

and we can write (31) as

$$\Re_2^t(\varphi_2) = X(2)D(2)X(2)^{-1}. \tag{35}$$

From (35), the transpose of $X(2)D(2)X(2)^{-1}$ is equivalent to multiplication by the Gauss polynomial φ_2 modulo $g_2(u)$.

15.7. <u>An Example</u> $N = 3^3$

Take $z = 2$ as a generator of the unit group U of $Z/27$. First consider the set of primitive multiplicative characters of $Z/27$

$$V(3) = \{x_k : 0 \le k < 18, \ (3,k) = 1\}. \tag{1}$$

Set

$$E^{\times}(3) = \{e_{z^k} : 0 \le k < 18\}. \tag{2}$$

A rational basis

$$R(3) = \{r_l : 0 \le l < 12\}, \tag{3}$$

for $ln(V(3))$ is defined by the matrix formula

$$R(3) = L(3)E^{\times}(3), \tag{4}$$

where

$$L(3) = \left[I_{12} \,\middle|\, -1_2 \otimes I_6\right]. \tag{5}$$

In fact, direct computation shows that

$$V(3) = X(3)R(3), \tag{6}$$

where

$$X(3) = \left[v^{kl}\right]_{0 \le k < 18,\,(3,k)=1,\ 0 \le l < 12}, \tag{7}$$

and $v = exp(2\pi i/18)$. The non-singularity of the Van der Monde matrix $X(3)$ implies $R(3)$ is a rational basis of $Ln(V(3))$.

Consider now the sets

$$V_2 = \{x_3, x_6, x_{12}, x_{15}\}, \tag{8}$$

$$V_2' = \{g_3, g_6, g_{12}, g_{15}\}. \tag{9}$$

To be explicit, we repeat the tables describing V_2 and V_2'.

Table 1.

	1	2	4	8	16	5
x_3	1	u	u^2	u^3	u^4	u^5
x_6	1	u^2	u^4	u^6	u^8	u^{10}
x_{12}	1	u^4	u^8	u^{12}	u^{16}	u^{20}
x_{15}	1	u^5	u^{10}	u^{15}	u^{20}	u^{25}

Table 2.

	3	6	12	24	21	15
g_3	1	u	u^2	u^3	u^4	u^5
g_6	1	u^2	u^4	u^6	u^8	u^{10}
g_{12}	1	u^4	u^8	u^{12}	u^{16}	u^{20}
g_{15}	1	u^5	u^{10}	u^{15}	u^{20}	u^{25}

The multiplicative characters in V_2 are defined on the rest of U by the periodic condition. The functions in V_2' vanish off of the set $3U$. We see from table 1 that the set V_2 can be constructed by periodizing mod 9 the primitive multiplicative characters $V(2)$ of $Z/9$ and we see from table 2 that the set V_2' can be constructed by decimating to $3U$ the set of functions $V(2)$. As a result, the set R_2, formed by periodizing mod 9 the rational basis $R(2)$ of $Ln(V(2))$, is a rational basis of $Ln(V_2)$. The set S_2, formed by decimating to $3U$ to rational basis $R(2)$, is a rational basis $Ln(V_2')$. Using (4.5),

$$V_2 = X(2)R_2 \tag{10}$$

$$V_2' = X(2)S_2. \tag{11}$$

Form the rational basis of $L(Z/27)$

$$f_0, f_1, f_2 \; ; \; x_0, g_0 \; ; \; x_9, g_9 \; ; \; R_2, \, S_2, \, R(3), \tag{12}$$

where

$$f_0 = e_0 + e_9 + e_{18}, \tag{13}$$

$$f_1 = e_3 + e_{12} + e_{21}, \tag{14}$$

$$f_2 = e_6 + e_{15} + e_{24}. \tag{15}$$

The action of the FT F of $Z/27$ will be described relative to the rational basis (12). We will do this in a sequence of steps.

1. $Ln(f_0, f_1, f_2)$

This is the space described in chapter 14 as $L(1,2)$ where we showed that the matrix of F relative to the basis $\{f_0, f_1, f_2\}$ is

$$3 \, F(3). \tag{16}$$

2. $Ln(x_0, g_0)$

Applying theorem 3 of section 4, chapter 14, the matrix of F relative to the basis $\{x_0, g_0\}$ is

$$-3 \begin{pmatrix} 0 & 1 \\ 1 & 0 \end{pmatrix}. \tag{17}$$

3. $Ln(x_9, g_0)$

Applying theorem 2 of section 4, chapter 14, the matrix of F relative to the basis $\{x_9, g_0\}$ is

$$\sqrt{3} \, i \begin{pmatrix} 0 & 1 \\ 9 & 0 \end{pmatrix}. \tag{18}$$

4. $Ln(R_2, S_2)$

First the matrix of F relative to the basis $V_2 \cup V_2'$ is the skew-diagonal matrix

$$G_2 = \begin{pmatrix} 0 & G(2) \\ 3 \, G(2) & 0 \end{pmatrix}. \tag{19}$$

where $G(2)$ is the matrix given in (18) of section 3. By the change of basis formulas (10) and (11), the matrix of F relative to the basis $R_2 \cup S_2$ is

$$X^{(2)}(2)G_2 X^{(2)}(2)^{-1}, \tag{20}$$

where

$$X^{(2)}(2) = X(2) \oplus X(2). \tag{21}$$

5. $\underline{Ln(R(3))}$

Set

$$G(3) = \begin{bmatrix} & & & G_3(x_{17}) \\ & & \bullet & \\ & & \bullet & \\ & G_3(x_j) & & \\ & \bullet & & \\ \bullet & & & \\ G_3(x_1) & & & \end{bmatrix}_{0 \le k < 18,\ (3,k)=1} \tag{22}$$

The matrix of F relative to the basis $V(3)$ is $G(3)$. By the change of basis formula (6), the matrix of F relative to the basis $R(3)$ is

$$X(3)\, G(3)\, X(3)^{-1}. \tag{23}$$

Putting these results together, we have that the matrix of F relative to the basis (12) is

$$F = 3\, F(3) \oplus (-3) \begin{pmatrix} 0 & 1 \\ 3 & 0 \end{pmatrix} \otimes \sqrt{3i} \begin{pmatrix} 0 & 1 \\ 9 & 0 \end{pmatrix}$$

$$\oplus\, [X^{(2)}(2)\, G_2(2)\, X^{(2)}(2)] \oplus [X(3)\, G(3)\, X(3)^{-1}]. \tag{24}$$

[References]

[1] Auslander, L. Feig, E. and Winograd, S. "The Multiplicative Complexity of the Discrete Fourier Transform", *Adv. in Appl. Math.* 5 (1984), 31-55.

[2] Tolimieri, R. "Multiplicative Characters and the Discrete Fourier Transform", *Adv. in Appl. Math.* 7 (1986), 344-380.

[3] Rader, C. "Discrete Fourier Transforms When the Number of Data Samples is Prime", *Proc. of IEEE*, 56 (1968), 1107-1108.

[4] Winograd, S. *Arithmetic Complexity of Computations*, CBMS Regional Conf. Ser. in Math. Vol. 33, Soc. Indus. Appl. Math., Philadelphia, 1980.

[5] Tolimieri, R. "The Construction of Orthogonal Basis Diagonalizing the Discrete Fourier Transform", *Adv. in Appl. Math.* 5 (1984), 56-86.

INDEX